Florida's Geologic Atlas:
A guide to county geologic maps

By

William A. Szary

Copyright 2021. Earth2Energy Educational Publishing.
All rights reserved.

Library of Congress Cataloging in Publication Data

Szary, William A.

Florida's Geologic Atlas: *A guide to county geologic maps*

Includes references and index.

ISBN 13: 9798723792111

Earth2Energy Educational Publishing
Port Richey FL 34668

Earth2Energy is a registered trademark

Table of Contents

Part I. Introduction 5
Florida's Physiographic Provinces
Florida's Marine Shoreline and Fluvial Terraces
Fluvial (River) Terraces

Panhandle Florida Surficial Geology

Northern Peninsular Florida
Atlantic Coastal Geology
Gulf Coastal Geology

North Florida
Interior Highlands Geology

South Peninsular Florida Geology

Part II. Geologic Maps 36
Panhandle Florida Maps 37
Bay County
Calhoun County
Dixie County
Escambia County
Franklin County
Gadsden County
Gulf County
Hamilton County
Holmes County
Jackson County
Jefferson County
Lafayette County
Leon County
Liberty County
Madison County
Okaloosa County
Santa Rosa County
Suwannee County
Taylor County
Wakulla County
Walton County
Washington County

North Peninsular Florida Maps 60
Alachua
Baker
Bradford/Union
Brevard
Citrus
Clay
Columbia
Duval
Flagler
Gilchrist
Hernando
Hillsborough
Lake
Levy
Marion
Nassau
Orange
Osceola
Pasco
Pinellas
Polk
Putnam
St. Johns
Seminole
Sumter
Volusia

South Peninsular Florida Maps 95
Broward
Charlotte
Collier
Dade
DeSoto
Glades
Hardee
Hendry
Highlands
Indian River
Lee
Manatee
Martin
Monroe
Okeechobee
Palm Beach
St. Lucie
Sarasota

Part I. Introduction

Florida's Physiographic Provinces

Most of Florida's subdivisions are distinguished from each other by the presence of distinct changes in elevation, sedimentary formation type, and geographic setting. There are approximately three major subdivisions that are recognized in Florida: the Gulf Coastal Lowlands, the Atlantic Coastal Lowlands, and the Interior Highlands.

Florida's landscapes are geologically young when compared to other state's geology. Exposure to subaerial erosion did not persist until the end of the Miocene Epoch, 5 million years ago. Erosional activities peaked during the Pliocene Epoch continuing into the Pleistocene and Recent (Holocene) Epochs. Physiographic provinces help identify the various types of landforms present throughout the state.

At least five different types of landscapes are directly related to the decomposition and disintegration of the underlying rock and sedimentary formations. These landscapes can be categorized as flat, monotonous plains which demonstrate subtle changes in elevation.

Gently undulating broad plains are generally flat for the most part but are occasionally interrupted by shallow depressions. Gently rolling hills are low relief in elevation. Moderate rolling hills occupy the higher elevations of the landscape, and steeply rolling hills consist of high relief elevations.

Relationships between certain landscapes and the underlying geology reveal patterns that consistently repeat throughout Florida (Upchurch and Randazzo, 1997). Interior broad flatlands positioned at higher elevations are supported by underlying clays which are impermeable to percolating water. Younger sand deposits rest on top of these clay units.

Florida's Marine Shoreline and Fluvial Terraces

There are five marine shoreline terraces that are recognized in Florida. Marine terraces represent former wave cut shorelines that were left behind when seas advanced and retreated from Florida during interglacial cycles. The terraces are believed to consist of sediments deposited under marine conditions which reached a maximum of 150 feet above present day mean sea level.

Terraces occurring above 150 feet are believed to consist of subaerial fluvial (river) terraces formed when river channels aggraded during sea level advances. River channel erosion occurred when sea levels retreated across the landscape.

Figure 1. Florida is subdivided into many physiographic regions. The physiographic map was modified by adding colors to each province from the map contained in Randazzo and Jones, 1997, Figure 1.12. The red line marks the extent of the Cody Scarp in Panhandle and Peninsular Florida.

Florida's Coastal Lowlands merge uplands with the marine environment. This zone parallels the coastline riming the entire Florida Peninsula and southern Panhandle regions, extending 30 to 50 miles inland. Lowlands are very sensitive to sea level and land use changes.

William A. White (1970) recognized at least three terraces occupying the coastal lowlands. The Silver Bluff shoreline (0 to 10 ft above mean sea level), Pamlico shoreline (10 to 25 feet above mean sea level), and Talbot shoreline (25 to 42 feet above mean sea level) are those terraces considered to represent the coastal lowlands. Healy (1979) assembled a map showing the boundaries of each recognizable terrace found within the state. The map has since been updated (**Figure 1**). The terrace map is very similar in arrangement to the physiographic map presented in **Figure 2**.

Changes in coastal landforms occur throughout the Gulf Coast region from the Panhandle through the Big Bend region into the central and southern parts of the Peninsular coastal region.

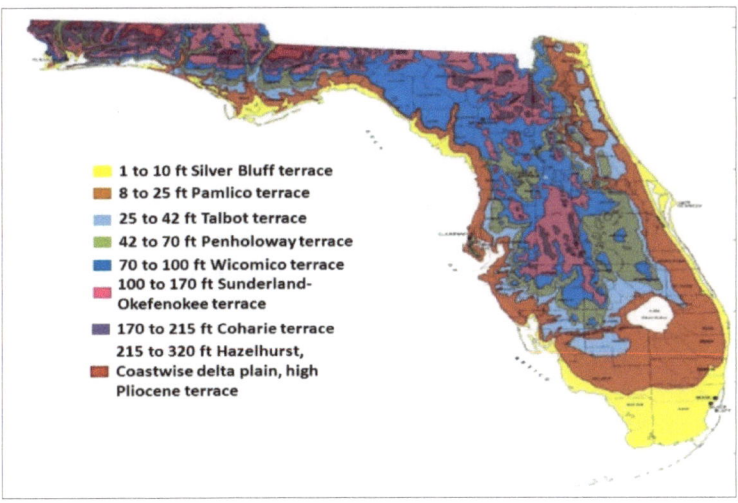

Figure 2. Florida's marine shoreline terraces covering the state when seas advanced and retreated throughout the Cenozoic Era (Healy, 1979).

Fluvial (River) Terraces

The high terraces mapped above 150 feet relative to present day sea level are considered to consist of river origin sediments as opposed to marine origin for the lower terraces. River deposition occurred during interglacial periods following the Kansan interglacial stage between 900 and 550 thousand years ago. Since the depositional history of the formation plays a significant part in recognizing the high terrace in Panhandle and Peninsular Florida, a discussion of the origin of the formation is expanded in the following section.

Panhandle Florida Surficial Geology

Throughout the Western Panhandle region, the **Coharie-Hazelhurst** terrace is severely dissected. Terrace remnants terminate on the gulfward side at an elevation of 200 feet above present day mean sea level. The seaward edge lies 50 feet above the Okeefenokee terrace. Above 150 feet, the Citronelle formation rests on top of Pleistocene age sea level terraces.

Table 1. Summary of Florida's Recognized Terraces

Terrace	Type	Elevation above mean sea level	Interglacial Period	Age
Silver Bluff	marine	1 to 10 ft	Mid-Wisconsin retreat	15,000 yrs ago
Pamlico	marine	8 to 25 ft	Mid Wisconsin retreat	15,000 yrs ago
Talbot	marine	25 to 42 ft	Sangamon retreat	400 to 150 thousand yrs ago
Penholoway	marine	42 to 70 ft	Sangoman retreat	400 to 150 thousand yrs ago
Wicomico	Marine	70 to 100 ft	Sangoman retreat	400 to 150 thousand yrs ago
Sunderland-Okeefenokee	fluvial	100 to 170 ft	Yarmouth retreat	900 to 550 thousand yrs ago
Coharie	fluvial	170 to 215 ft	Yarmouth retreat	900 to 550 thousand yrs ago

Citronelle sediments accumulated from several early rivers which emptied into the Gulf of Mexico during interglacial sea level rise. Speculation concerning the origin of Citronelle sediments suggests deposition resulted from the aggradation by rivers during the Pliocene or during the rise of seas due to interglacial period ice cap melting in the northern regions of the continent.

When the rise in sea level occurred during the Aftonian interglacial stage (post Nebraskan glacial stage-1.7 to 1.4 million years ago), the Citronelle formation was partially submerged as a platform. During the Kansan stage (1.4 million years to 900 thousand years ago), rivers cut gaps into the high terrace.

Following the Kansan stage, the Yarmouth interglacial period seas rose to 150 feet above present day levels. River transported sediments reached the sea, distributed coastwise by oceanic waves and longshore currents. Following sea level retreat, Citronelle deposits were left behind.

East of the Apalachicola River, erosion removed most of the once continuous Coharie-Hazelhurst terrace in the Florida Panhandle. Elevations rise to about 280 feet in this part of the region. No clearly defined scarp exists to identify this terrace from the underlying marine terraces.

The **Okeefenokee** shoreline was eroded between the Apalachicola River and Walton County. The terrace is well defined between Walton and Okaloosa Counties, trending in an arcuate shaped pattern. A set of islands enclosed the intracoastal waters of southeast Georgia to the south. In northern Hamilton County, the intracoastal terrace of the Okeefenokee sea entered Florida, extending into the Tallahassee vicinity. Between Tallahassee and the Apalachicola River, the shoreline was straightened with several submerged offshore sand bars.

The **Penholoway** terrace is dissected by erosion extending northward along major rivers including the Chattahoochee, Chocktawhatchee, Blackwater, and Escambia Rivers. The Walton County geologic map suggests the Choctawatchee River Penholoway terrace is underlain by Alum Bluff and Citronelle Formations lying in contact with Quaternary alluvial floodplain sediments. Penholoway shorelines were more recently recognized, occupying westerly trends from Tallahassee into the south central part of Leon County, curving in an arcuate pattern paralleling the Wicomico shoreline to the Apalachicola River. The river below Seminole Lake cut into the shoreline exposing the Pamlico terrace deposits along its channel, banks, and floodplain.

In Western Florida, the **Wicomico** shoreline is irregularly shaped. The terrace formed a gentle arc north of Choctawhatchee Bay paralleling the Okeefenokee shoreline. During late Wicomico time, well defined sand bars were constructed east of the Ochlockonee River. Example Wicomico landscapes can be viewed west of the Apalachicola River.

The **Talbot** terrace is restricted in extent to the western and central Panhandle regions rimming the interior parts of the coastal plain region extending from southern Leon County west, thinning in width near Chocktawatchee Bay to Pensacola Bay. Talbot shorelines parallel the interior side of the Pamlico shoreline terrace.

West of the Apalachicola River, the **Pamlico** shoreline extended across the Wicomico terrace to a cape located at Panama City, continuing westward along the coastal shoreline along a series of sand bars coincident with the present day shoreline.

West Bay and Choctawhatchee Bay were larger than at present. West of Choctawhatchee Bay, the Pamlico sand bars were present along the north side of the Recent Santa Rosa Sound. Intracoastal water was present in the East Bay inlet and enlarged Penscola Bay. A series of sand bars and spits were positioned west of Pensacola Bay.

The Pamlico Apalachie Bay extended from the Gulf Coast into southern Leon County. The shoreline trended southward, southwest of Tallahassee along the Middle Miocene scarp west of the Ocala Platform. The shoreline of the western bay fell inline with the west side of the Recent bay shoreline. Near Medart (Wakulla County), a hooked cape cut across the Wicomico terrace in a general westerly trend.

Tates Hell Swamp (Franklin County) occupied a group of small arcuate shaped islands, suggesting the Apalachicola River emptied at this location during Pamlico time. It is unclear whether or not the sand bars along the north side of St. George Sound formed during Pamlico time.

The **Silver Bluff** terrace was located along the present day intracoastal and coastal shorelines. A large bay existed in southern Gulf and west Franklin Counties. The Apalachicola River shifted its channel, emptying into a depression after silting at the river mouth closed the channel on the east during Pamlico time. East of St. Joseph Bay, well defined sand bars built up, curving in an opposing direction to the Recent (Holocene) St. Joseph Spit. The present day bays, including the St. Andrews, West Chocktawhatchee, and Pensacola Bays were larger during Silver Bluff time between 6000 and 4000 years ago. Present day sea levels represent a retreat from the Silver Bluff terrace.

The remainder of inland Panhandle and Peninsular Florida makes up the Interior Highlands Province. The region is underlain by five major terraces: the Penholoway (42 to 70 feet), Wicomico (70 to 100 ft), and Okeefenokee-Sunderland (100 to 150 feet). The Okeefenokee terrace serves as a transitional slope between ridge crests and coastal lowlands. Higher ridge crests are occupied by the Coharie and Hazelhurst terraces which were formed by fluvial processes, not marine processes.

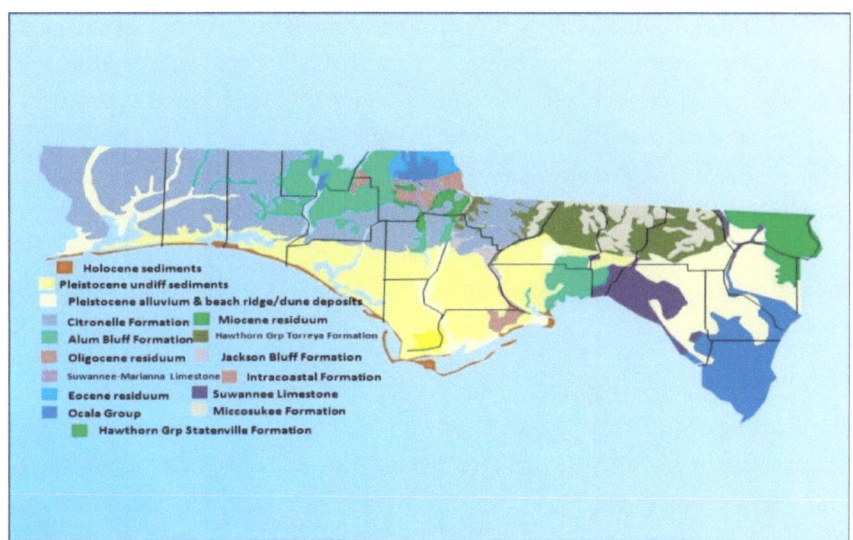

Figure 3. Portion of the Florida State Geologic Map showing the Panhandle regional geology which is exposed at the surface. Source: Redrawn from the USGS Florida state geologic map.

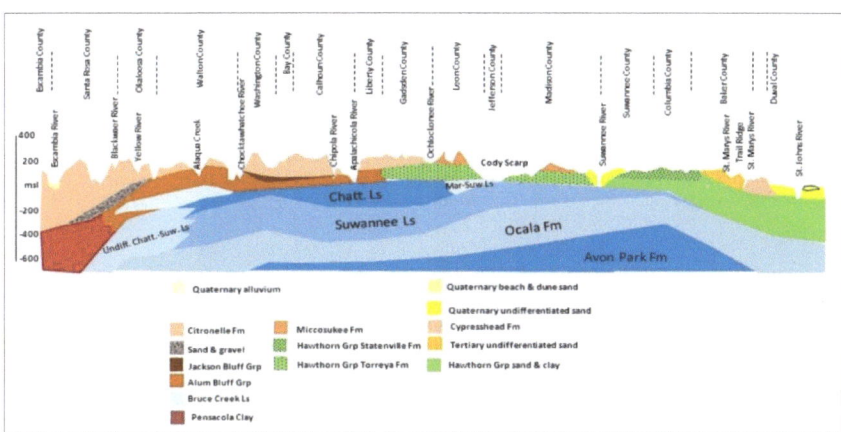

Figure 4. Cross section across Panhandle Florida following FL 4 in Escambia, Santa Rosa, and Okaloosa Counties, then following I10 to the North Florida east Atlantic Coast. Depths were obtained from well log interpretation by the USGS. Source: Redrawn from the Florida Panhandle cross section provided on the USGS State Geologic Map website.

The higher terraces form rolling hills, carved by modern day tributary and river downcutting which eroded into Miocene, Pliocene, and Pleistocene sedimentary formations. In Panhandle Florida, the Marianna Lowlands separates the Northern Highlands from the Western Highlands, residing on top of the Sunderland-Okeefenokee terrace. Stream erosion combined with limestone solutioning reduced once continuous hills to elevations lower than the New Hope and Grand Ridges. Both ridges border the southern limits of the lowland region occurring at a maximum elevation of 320 feet above mean sea level.

White (1970) suggested the Marianna Lowland resulted from Pleistocene downcutting of the Apalachicola River when sea levels gradually receded. The Chattahoochee and Chipola Rivers both downcut in an attempt to keep pace with the lowering of the Apalachicola River in the eastern and central parts of the lowlands. Subsequently, the Chocktawhatchee River downcut in response to retreating sea levels along the western side of the lowlands. Regional ground waters were lowered in Jackson and Holmes Counties. This left previously dissolved limestone cavities unsupported, exposed to air pockets causing the overlying sediments to collapse into the void spaces. Solution valleys expanded between clay supported ridges and hills.

The Tallahassee Hills, beginning at the Apalachicola River Valley, extend the Western Highlands into Madison County where they merge with the Northern Highlands.

Northern Peninsular Florida
Atlantic Coastal Geology

East Coast **Silver Bluff** terrace deposits appear in the form of sand bars preserved south of the middle Mantanzas River region, in the northern and southern parts of Ormond Beach, specifically in the Port Orange and New Smyrna areas. The Silver Bluff also occupies intracoastal lagoons between Mantanzas and New Smyrna in the form of narrow unnamed depressions. Merritt Island in Brevard County, and the Indian River occur at an elevation of 10 above mean sea level. The Indian River functioned as a lagoon during Silver Bluff time.

Seas advanced across the coastline of Florida between 25 to 35 feet above the present day mean sea level following the retreat of the Wicomico sea leaving behind the **Pamlico Terrace** occupying terrace elevations between 8 and 25 feet above mean sea level.

Sediments appear from St. Marys River into central Flagler County. Narrow sand bars extended into southern Brevard County. The shoreline terrace occupies a slightly wider belt which occurs 40 miles inland southward into Martin County. The belt widens to 100 miles near Lake Okeechobee into Broward County, curving into the Everglades region connecting with the Gulf Coast Lowlands. Pine tree scrubs occupy the terraced area except in the Big Cypress and Everglades region.

The **Talbot** terrace overlies the Pamlico terrace from Nassau to Volusia County. The terrace marks the interior boundary of the Pamlico surface. Rising seas responsible for depositing Talbot sediments flooded the St. Johns River floodplain. Higher sea levels prevented sedimentation from occurring within the Eastern Valley. When seas retreated, sedimentation accumulated to form the Atlantic Coastal Ridge and Center Park Ridge dune deposits.

In southern Brevard County, the shoreline of the lagoon continued south through Indian River and St. Lucie Counties extending through a sharp cape in southwestern Martin County marking the southern tip of the Orlando ridge. The shoreline continued around the north end of Lake Okeechobee trending southwest to the Caloosahatchee River then northwest to Sarasota Bay. In the Sarasota Bay area, the shoreline is marked by a low scarp cut into the Wicomico terrace. A large, low island existed south of the Caloosahatchee River.

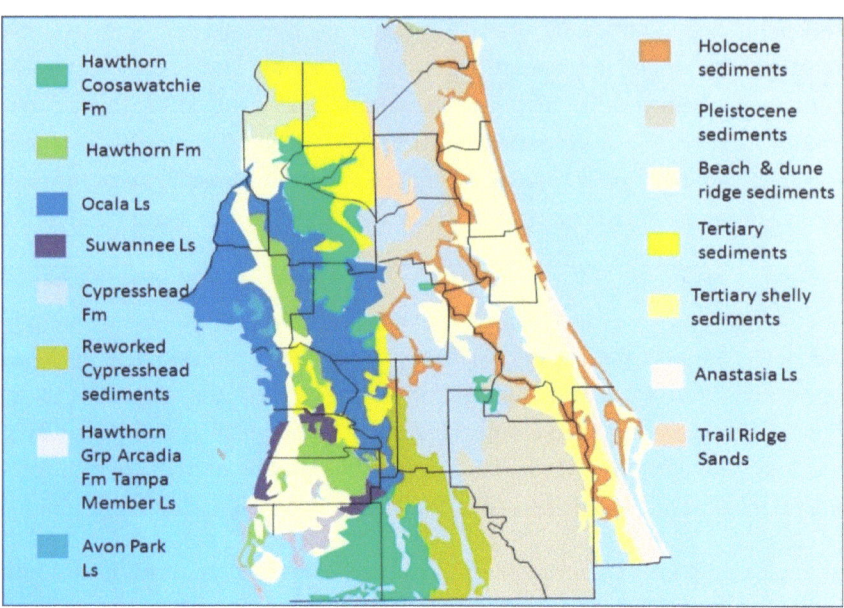

Figure 5. Portion of the Northern Florida State Geologic Map showing formations exposed at the surface. Source: Redrawn from the USGS State geologic map.

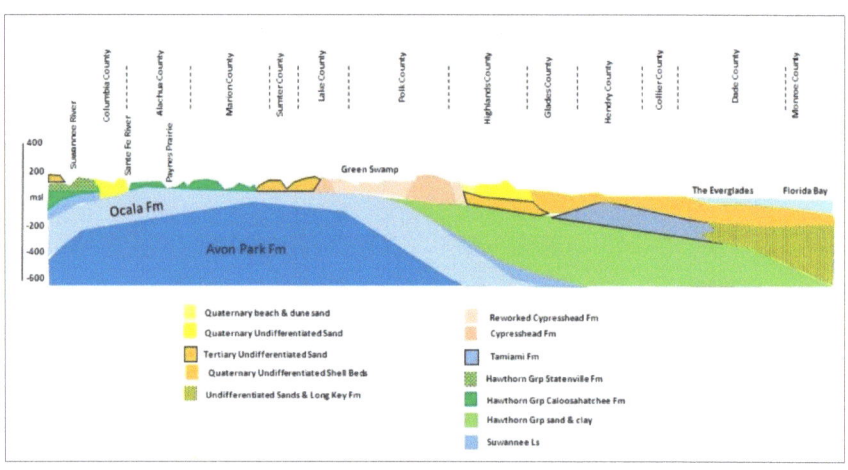

Figure 6. Cross section through the central Florida Peninsular region from the Georgia state line south into the Everglades region. Source: Redrawn from the cross section provided on the Florida State Map posted on the USGS State Map Website.

St. Johns County geology represents retreating shorelines in response to the withdrawing Atlantic Ocean towards the east. When seas retreated, undifferentiated sands were transported and spread across the western and central part of the county covering the landscape. Beach and dune sands accumulated from wind blown sands piling up along the retreating beach face in the central southeastern part of the county.

Rivers, keeping pace with the withdrawing Atlantic Ocean deposited alluvium in lagoons trapped by sand dune and beaches in the northeastern part of the county. Anastasia limestone formed in offshore bars which were covered over by beach sands during sea retreat. Holocene lagoons formed closer to the present day shoreline when tides moved between barrier islands forming directly adjacent to the ocean front.

The St. Johns River was discharging and transporting a greater volume of sediment during sea retreat, depositing a much larger floodplain area than when seas covered the eastern coastal plain lowlands. When seas fully retreated, floodplain sediments became exposed along the western region of the county.

The geological configuration of Volusia County is unique from the perspective of observing the process of sea advancing and retreating cycles and associated impacts to landscape development. During sea retreat from west to east across the county, sheet sands were deposited along with beach and dune sands, indicating former positions of relict shoreline stands.

The St. Johns River origin is the subject of debate. Earlier geologists think the river developed during Pamlico time as a lagoon depression. More recent theories suggests the river developed during sea retreat when the build out of a beach and dune ridge plain occurred. The river channel and floodplain was determined in part by the location of swales between relict beach dune ridges. The northern offset segment was probably formed earlier before Pamlico time, during the late Tertiary to Early Pleistocene Epoch (Randazzo & Jones, 1997).

Eastward of the St. Johns River floodplain deposits, remnants of the lagoon were later eroded exposing the underlying Cypresshead Formation red orange channel gravels and sands, and floodplain clays. Dune sands continue to occupy the upper Cypresshead sediments in the northwestern and southwestern parts of the county. The St. Johns River valley was a large intracoastal bay occupied by several large islands.

The bay was part of the open ocean south of Brevard County. Beach and dune sand deposits formed along the shoreline in the central eastern part of the county from continued sea retreat to the east.

While seas covered over much of the county, Anastasia limestone was accumulating in offshore bars which became exposed when sea retreat reached its present day position. Exposures occur along present day depressions which formed by lagoons occupying beach and dune ridges left behind when seas retreated to the east.

The present day barrier beach islands located along the Atlantic Ocean shoreline, and the present day intracoastal waterway behind the barrier beach fronts represent the same pattern of lagoons that appeared when sea retreat was active duing Quaternary time.

Gulf Coastal Geology

In the Big Bend Gulf Coast region, coastal swamplands dominate. Further south in Pinellas County, the coast returns back to barrier beaches, sand bars, lagoons, and isolated interior swamps extending into South Florida. In the Peninsular Florida region, the **Silver Bluff** terrace is replaced by the Pamlico terrace which maintains a relatively uniform width. Erosion from the latest rise in sea level removed the Silver Bluff terrace between Citrus and northern Charlotte County leaving Pamlico terrace sediments along the shoreline edge. Between Taylor and Pinellas Counties, sand supplies are deficient because the offshore slope is floored by a gently sloping limestone platform.

The harder rock material acts as a wave break far offshore, buffering wave energy before the energy can reach the shoreline zone. A source of resistant sand is not available for distribution along the shoreline. Instead, swamps trap sands, silts, and muds transported by rivers from eroding inland areas.

In the Tampa Bay region, the area was occupied by a large open estuary with several islands present. The largest island was located in St. Petersburg. Two other islands existed in Seminole and Tarpon Springs (Pinellas County).

During **Pamlico** time, the shoreline was muddy or had small amounts of sand from older terraces mixed in between Pasco and Wakulla Counties. Early Tertiary limestones were exposed by tilting of the Ocala Platform caused by differential settling. The Pamlico shoreline extended across Tampa Bay along the south shore of Hillsborough County into Hillsborough and McKay Bays east of the City of Tampa.

Within the interior parts of the Gulf Lowlands, the **Talbot** terrace occupies elevations ranging between 25 and 42 feet above mean sea level. Limestone either appears at the surface, or is buried by thin sediment layers. The Talbot terrace was entirely removed by erosion from the Big Bend region between Jefferson and Hernando Counties.

Taylor County represents a coastal carbonate platform exposed by erosion of overlying Miocene and Quaternary aged sediments. In the southern part of the county, Middle Eocene Ocala Group limestones are exposed where Suwannee Limestones were eroded when seas retreated after the Oligocene Period.

During the Pleistocene sea rise, Quaternary undifferentiated sands were distributed across the county, later removed by withdrawing sea levels to the west and by tributary and river erosion cutting into the more resistant Suwannee Limestone. Miocene aged formations were removed by erosion following sea retreat before the Pliocene Epoch began. Quaternary sands remained in the eastern half of the county, currently being dissected by tributary creek and river erosion.

South of Perry, the Suwannee Limestone was exposed to ground water lowering following the Pleistocene Epoch sea withdrawal, causing the surface of the Suwannee Limestone to develop solution sinkholes.

During sea retreat, a linear stand of beach dune sands were stranded north of CR 361, and parallel to CR 361 southward to Steinhatchee, FL. Holocene aged estuaries mark the edge of the Florida Platform along the Gulf of Mexico, occupied by mangrove swamps and marshlands. The coastline represents a low energy coastline where longshore currents are absent. High energy coastlines are identified by the presence of barrier beaches and sand spits.

The Pamlico shoreline covered the limestone part of western Taylor County, and the Wicomico shoreline covered the Quaternary undifferentiated sand unit. The Penholoway shoreline sediments were eroded from the landscape when the Pamlico terrace was deposited.

During the Middle Eocene Epoch, Levy County was submerged beneath a carbonate platform where the Avon Park Formation limestones accumulated. During the Upper Eocene, the platform was emergent when seas retreated. Erosion removed the upper portion of the Avon Park before the Ocala Group limestones accumulated in advancing seas again during the uppermost Eocene Epoch.

Another period of erosion occurred from retreating sea levels which removed portions of the Ocala Group from above the Avon Park Formation in the south central part of the county. Shallow marine clastics and deltaic deposits formed the Hawthorn Group which was weathered from exposure during sea level retreat to the west during the Miocene Epoch. Sea retreat during Pliocene and Quaternary time continued weathering the Hawthorn Group sediments while depositing sands in the central and eastern part of the county on top of exposed Ocala limestone.

Lowering sea levels also lowered ground water levels which began karstic sinkhole activity around the western edge of the Quaternary sands and along the eastern edge of the Hawthorn Group weathered sediments.

Gilchrist County is underlain by the Ocala Group limestone platform exposed when the platform became tilted from differential settling. The younger Suwannee limestone was eroded from the landscape during the tilting process. When the Gulf of Mexico retreated from the county, undifferentiated sands were left behind by sea distribution across the central part of the county.

During the Miocene Epoch, clastic marine sediments built out deltas ending up along the shallow shelf formed by the Eocene limestone platform. When seas retreated, the sediments were weathered and eroded from the landscape, leaving behind a section exposed on top of the Ocala limestone in the southeastern part of the county. The exposure belongs to the weathered Hawthorn Group sediments.

Following retreat of the Gulf of Mexico, ground water levels dropped in response to the retreating sea level. Karstic sinkholes began collapsing the undifferentiated sands into caves and caverns formed in the underlying Ocala Group along the central and eastern portions of the county.

Pasco County geology exposes the Upper Eocene Ocala Group limestone along the eastern boundary with Sumter County when Pasco County was covered by warm shallow seas. Seas withdrew for a brief time at the end of the Eocene Epoch but returned again during the Oligocene Epoch, depositing the Suwannee Limestone on top of the Ocala limestone, presently exposed along the southeastern corner of the county, and along the western coastline. Salt water estuaries and marsh systems cover the Suwannee limestone along the Gulf of Mexico coastline.

Seas withdrew again, but returned during the Miocene Epoch when the Arcadia Formation Tampa Member limestone was deposited on top of the Suwannee Limestone. Tampa Limestone is exposed in the central south part of the county. Miocene Hawthorn Group clays accumulated in an enlarged Tampa Bay which covered the county after seas began to retreat to the west. Hawthorn clays support steeply rolling hills in the central east part of the county around Dade City, and remnant clays occupy the central western county.

When the old Tampa Bay receded to the west, undifferentiated sands were left covering most of the central part of the county with dunes left behind along the western and northwestern coastlines marking the position of the Gulf at the end of the Pleistocene Epoch. Sinkhole lakes formed when seas retreated back into the present day Gulf of Mexico, lowering the ground water beneath the county. Caves, caverns, and fractures developed in the limestone beneath the Hawthorn clays and undifferentiated sands, collapsing the younger formation units into sinkholes. This process resulted in the steep hills in the central east part of the county, and shallow sinkhole depressions in the central and western parts of the county.

Marine shoreline terraces are difficult to recognize because they appear as subtle changes in the landscape supported by the Quaternary undifferentiated sand unit. The coastline extending about 10 miles inland from the coast belongs to the Pamlico terrace.

Between 10 and 15 miles inland from the coast, the Talbot terrace appears as a slight increase in elevation marking the contact with the Penholoway terrace. In the central part of the county, at the contact between the Hawthorn Group clay and Quaternary undifferentiated sand unit, the Wicomico terrace appears. Much of the Hawthorn Group formation itself supports the Sunderalnd-Coharie fluvial terrace of the Brooksville Ridge.

Between Dade City and Brooksville, high hills stood as islands during the Okeefenokee sea stand. The hills are composed of Hawthorn formation beds which were leached in place. In Hernando, Citrus, Levy, and Marion Counties, small hills occupy elevations greater than 150 feet above present day mean sea level. Islands were also present.

In Pasco County, a strait existed across the site of Lake Butler, and a spit curved around the north end of the lake. The spit was part of the Anclote Keys, now exposed off the Gulf Coast, the last exposed sandy feature occurring between south Pasco County around the Big Bend region into Franklin County.

Most of the Pinellas County limestone formations were covered over when seas advanced across the peninsular region about 1 million years ago. A small portion of the underlying Hawthorn Group Arcadia Formation Tampa Member limestone was left exposed from later erosion resulting from the Anclote River discharge into the Gulf of Mexico, north of Lake Tarpon. Small remnants of dune sands remain on top of the limestone in the Sunset Hills area west of Tarpon Springs. During the Lower Miocene Epoch, a closed basin depositional environment existed which allowed sand, silts, and clays to accumulate along with carbonate deposition.

Phosphate minerals formed when cold water upwelled from deeper parts of the Gulf into shallower parts, promoting precipitation of carbonate fluorapatite found in the Tampa Member of the Arcadia Formation. The Hawthorn Group sandy clays form the more resistant uplands along the northwest and central parts of the county.

About 15,000 years ago, seas retreated to the west leaving behind undifferentiated shell beds, and sands covering over the Tampa Member basin limestone. Barrier beach islands along the southwestern central and southern coast formed when Holocene aged lagoons and estuaries formed beach fronts, cutting off the intracoastal lagoon from the Gulf of Mexico. Anclote Key, Honeymoon Island, and DeSoto Park are remnants of the lagoonal beach and estuary deposits left behind to form small nearshore islands along the western coast.

To the east of Pinellas County, Hillsborough County exposes Oligocene Suwannee Limestone in the northeastern corner of the county along the Hillsborough River representing a carbonate platform. Older carbonate rocks belonging to the Ocala Group and Avon Park Formation were buried beneath Suwannee carbonates. Due to the urbanized nature of Pinellas County, natural photographs representing geologic conditions are rare. Most of the formations shown on the geologic map are covered over by pavement.

During the Miocene Epoch, Tampa Bay covered much of Hillsborough County, depositing mixtures of clays and limestone belonging to the Hawthorn Group Arcadia Formation Tampa Limestone. Tampa Limestone is exposed along the Upper Tampa Bay around the City of Tampa peninsula, and along the Hillsborough River trending to the northeast in the north central part of the county. The Hawthorn Group Peace River Formation Bone Valley Member was deposited on top of the Tampa Limestone covering most of the eastern half of the county.

Seas retreated from the county leaving behind undifferentiated shell beds along the southwestern coastline along with estuary marsh systems. Erosion from the Hillsborough and Palm Rivers cut through Quaternary sands exposing the underlying Oligocence and Miocene limestones. Coastal erosion around Tampa Bay also exposed Miocene Tampa Member limestones along the southwestern edges of the county, and around the City of Tampa peninsular area. The Alafia River removed most of the Quaternary sands from the Peace River Formation Bone Valley Member in the eastern, central, and northeastern parts of the county. Undifferentiated shell beds occupy marsh type estuaries along the Gulf of Mexico in the southwestern part of the county in the vicinity of Ruskin.

North Florida
Interior Highlands Geology

The North Florida interior highlands region consists of a mixture of landscape types, most notably flat level plains interrupted by moderately to steeply rolling hills. Traveling north or south along Interstate 75 provides a unique cross section of the hills and valleys shaped by tributary stream and river erosion, and landscape collapsing formed by dissolving limestone appearing at or near the surface or buried deeply beneath clays. The Northern Highlands of Peninsular Florida consists of a set of gently sloping plateaus which transition the upper highlands into coastal lowlands. These plateau surfaces were once continuous but have since undergone extensive erosion and collapse from subsurface limestone solutioning. The Northern Highlands are bounded to the south by the Florahome Valley along the southern edge of Alachua and Putnam Counties.

The Central Florida interior highlands consist of a series of parallel north to south trending sand ridges interupted by parallel trending broad valleys. The ridges represent old beach sand dunes left behind when Pleistocene seas retreated from the interior about 15,000 years ago. Some of the higher ridges are held up by underlying clays and sandy clays formed by Miocene and Pliocene relict deltaic and estuarine sediments. Pleistocene sands were deposited on top by advancing and retreating seas.

Most valleys are formed by modern day collapse of underlying limestone formations. Broad shaped valleys are occupied by shallow lakes which rest on top of former sinkhole depressions. Deeper lakes are found on top of ridge crests. Both shallow and deep lakes are fed by pressurized ground water flowing upward from limestone aquifers.

The present day landscape is characterized by well drained, well developed dendritic (branching) stream patterns where thicker uniformly distributed sediments cover limestone. Where limestones are exposed close to the surface, trellis stream patterns trace limestone fractures and joints. Rolling hills and abrupt ridges mark former terrace surfaces bounded by stream channels or sinkhole rims. Broad shallow basins belong to water filled sinkhole scars. Where the interior highlands descend onto coastal lowlands, the contact forms the Cody Scarp. The scarp marks the boundary between the Western Highlands in Panhandle Florida and Coastal Lowlands, and the boundary between the Northern Highlands and Coastal Lowlands in Northern Peninsular Florida.

Okeefenokee (aka Sunderland) shoreline sediments were submerged during Trail Ridge deposition. The ridge widened in Clay and Bradford Counties to an enlarged pear shaped promontory. The southern end was probably an island of older terrain rather than a sand bar.

In southern Baker and northern Bradford Counties, a lagoon occupied the western Trail Ridge area, shallowing to the south. The Okeefenokee shoreline cut back westward into southeast Lake County, changing direction to the south along the east side of the Lake Wales Ridge to a position located south of Lake Child.

Three irregular, roughly parallel ridges stood as islands in the Okeefenokee sea in Polk and Highlands Counties in central Peninsular Florida. The Lake Wales Ridge was one of these ridges. Aprons of white sand extended outward from the actual shoreline for a short distance. The aprons were submarine extensions of beaches. White sands were the erosional remnants originating from the central highland ridges. Smaller promontories were the result of reduced marine abrasion to wave washed sand bars.

Irregular, thin white sand dunes are positioned between 150 and 225 feet above mean sea level. Greater than 150 feet, terrains are hilly with many round lakes. Lakes appear absent below 150 feet in elevation. Lakes that existed during Okeefenokee time were filled in by the sea. Lakes were thought to be the product of sinkhole activity. The central highlands are older than the shoreline terrace. The Okeefenokee terrace is well preserved in northern Florida, along both sides of the Lake Wales Ridge and the two smaller ridges which make up the central highlands in Peninsular Florida. The highest point of the ridge is 325 feet at the Bok Tower Gardens in Lake Wales, Florida.

The ridge appears to be a continuation of the northern Trail Ridge. Both ridges trend along the broadly arcuate Okeefenokee shoreline. Portions of the terrace were built up while other portions were eroded to conform with the alignments of both ridges.

Invading seas flooded valleys above weathered Miocene aged rocks. The Okeefenokee shoreline followed the east side of the Orlando Ridge in Marion, Lake, and Orange Counties. A small island occupied the ridge in northern Marion County. From southern Marion County, the shoreline extended into south central Orange County forming a prominent cape southeast of Orlando.

Wicomico shorelines are the least recognizable, suggesting sea level stands were relatively short in duration. Wicomico sands merge with the Trail Ridge, similar in character to the Okeefenokee scarp extending southward into Duval County. Well defined sand bars were present north and south of the St Marys River during late Wicomico time. In Clay, Putnam, and Alachua Counties, the shoreline is very irregular. The Orlando Ridge was a long narrow peninsula during late Wicomico time occupying southern Orange County. The ridge extended southward decreasing in elevation through eastern Osceola, northeastern Okeechobee, and western St. Lucie Counties into western Martin County. The lower south end of the ridge is probably younger than the northern counterpart. The ridge grew southward when the Wicomico sea retreated. Southward moving longshore currents redistributed Wicomico coastal sand bars along eastern Florida. This process continues today.

In Polk County, the Wicomico shoreline occupies the region between Lake Childs northward to Sebring. The shoreline is irregularly shaped around the Central Florida uplands. The terrace appears in the form of a low relief scarp cut into the older Okeefenokee terrace. Wicomico sediment appears along the west side of the Orlando Ridge from Lake Apopka into central Marion County. The terrace rejoined the eastern shoreline at this location. Large islands were present in Pasco, Hernando, Citrus, Sumter, and Marion Counties.

At Gainesville in Alachua County, Wicomico sands parallel the strike of Miocene beds around the Ocala Platform extending north to Tallahassee.

In the remainder of this chapter, the surface geology is described for those counties which intersect with I75 from Columbia County south to Pasco County.

The oldest rocks in Columbia County are exposed along the Sante Fe River which marks the southern boundary of the county. River dissection has cut into Quaternary and Tertiary aged undifferentiated sands, and Hawthorn Group clays to reveal the Ocala Group limestone. Tributary creek, stream, and river erosion removed most of the Tertiary aged undifferentiated sand from above the Hawthorn Group clays in the central part of the county producing hilly landscapes. Hills are covered by Tertiary and Quaternary undifferentiated sand units, exposed in the northern and southern parts of the county .

In the vicinity of Lake City, and north towards Interstate 10, the Hawthorn Group produces moderately to gently rolling hills often becoming flat. Three landscape domains can be recognized in the Lake City region: Lake City occupies the upland domain which consists of a flat platform underlain by Hawthorn Group clays and Tertiary undifferentiated sands uninterrupted by underlying limestone formations belonging to the Suwannee and Ocala Formations. Limestones were not exposed to karstic processes.

South of Lake City, a narrow band of Hawthorn Group clays are interrupted by karstic features which make up the scarp domain. Steeply rolling hills are separated by intervening valleys occupied by sinkhole lakes and ponds. A small section of Tertiary undifferentiated sand covers a localized region of Hawthorn Group sediments east and southeast of Lake City. The third domain belongs to the lowland domain covered over by Quaternary undifferentiated sands forming low hills interrupted by surface sinkholes present in the underlying limestone that appears closer to the surface than in the previous two domains. Limestone occurs near the surface, covered by a thin veneer of sand and clay which sink into limestone karstic features instead of collapsing into depressions.

The Ocala Group is the oldest limestone unit exposed in Alachua County. The Suwannee Limestone unit was eroded from the landscape when seas withdrew during the Upper Oligocene Epoch and the Ocala Platform began to settle differentially. Hawthorn Group sediments (including limestone) were deposited on top of the Ocala Group. Portions of the Hawthorn Group were weathered and eroded from the upper Ocala Group surface leaving behind a remnant patch in the southwestern corner of the county. Most of the central part of the county remained covered by Hawthorn Group clays which are responsbile for producing the steeply rolling hills along Interstate 75.

The hills are formed by internal drainage resulting from karstic processes occurring in the underlying Ocala Group limestone.

During Pliocene time, Tertiary undifferentiated sands and Cypresshead sandy clays covered Hawthorn Group sediments in the eastern central and far eastern regions of the county. The large open surface water feature in the southeastern edge of the county belongs to Paynes Prairie. The prairie developed when ground water receded allowing a former lake to drain into the Ocala limestone. Over many cycles, sinkhole collapsing occurred from weakened caves and cavern systems which progressively expanded the prairie basin.

In Alachua County, an island composed of Hawthorn Formation was surrounded by the Okeefenokee sea along with the south end of Trail Ridge. Other large islands existed in Baker, Columbia, Suwannee, and Hamilton Counties. These islands represented the upturned edge of the Hawthorn Formation around the east and north sides of the Ocala Platform.

Marion County landscapes are similar to Alachua County. Flat plains occur on Upper Eocene Ocala Group limestone in the southwestern and central parts of the county where it lies at or near the surface. Patches of Miocene Hawthorn Group clays rest on top of the Ocala limestone unit forming steep hillslopes which interupt flat plains in the western half of the county. Younger formations overlie the eastern half. Tertiary undifferentiated sands lie along the former Oklawaha River floodplain surface now occupied by extensive fluvial wetland marshes.

The river channel cuts into Holocene aged fluvial deposits transported southward from northern counties. These sediments rest on top of Pliocene aged Cypresshead Formation sandy clays which collapsed from dissolution of the underlying Ocala limestone, forming numerous lake depressions along the eastern edge of the river floodplain. In the eastern part of the county, Quaternary undifferentiated sands cover the Cypresshead sediments. Remnant weathered Hawthorn Group clays are exposed from sand erosion within the younger undifferentiated sand unit. Paynes Prairie occupies the central north open water feature.

Sumter County exposes the Upper Eocene Ocala Limestone throughout most of the central and western parts which make up flatlands. Surface sinkholes formed in the upper limestone surface due to exposure of limestone to infiltrating rainfall and high ground water table conditions which dissolves limestone.

When ground water levels fall, the limestone is no longer supported which begins the collapsing process due to the weakened rock structure. In the north central western part, the Ocala limestone was weathered into clays and sandy clays leaving behind residuum sediments covering part of the limestone formation.

In the central part of the county, Tertiary undifferentiated sands cover the limestone unit of which the limestone is partially exposed in the vicinity of Coleman, where erosion removed the sand unit. In the northeastern part, Pliocene Cypresshead Formation sandy clay is exposed. Sandy clays support hilly landscapes which are interupted by linear valleys which represent limestone collapsing occurring beneath clayey units.

In the easternmost part of the county, Cypresshead sediments were weathered, transported, and redeposited on top of the Ocala Limestone by sinkhole collapse and tributary drainage. Small patches of Quaternary dune sands occupy the upper Tertiary sands south of Coleman.

The oldest unit exposed in Lake County belongs to the Miocene Hawthorn Group sandy clays. Limited exposures appear in the central east part of the county typically associated with low lying depressional features where springs occur. Some river channels expose this unit as well. Hawthorn Group sediments occur as sand, silt, and clay deposited under marine conditions. Hawthorn sediments unconformably overlie Upper Eocene Ocala Group limestone. Pliocene Cypresshead sandy clays overlie the Hawthorn sediments unconformably.

Cypresshead Formation sands and clayey sands accumulated in a shallow, nearshore environment covered by Quaternary beach shoreline and dune sands, and Holocene aged fluvial sediments left behind when seas withdrew from the county. Quaternary undifferentiated sands and clayey sands cover Cypresshead sediments accumulated in marine settings although some erosion, transport, and redeposition occurred at the time of deposition.

Cypresshead sediments were also weathered, eroded, transported, and redeposited on top of the Cypresshead Formation in the southwestern part of the county at this time.

Quaternary dune sands are present in northeastern Lake County. Karst modification of the dune sands occurred during the Holocene. The St. Johns River floodplain and tributaries draining into the river deposited fluvial and lacustrine sediments consisting of sands, clays, silts, and organic matter.

The large surface lakes in the western part of the county developed by small sinkhole lakes which gradually expanded by collapse of the underlying limestone cave and cavern system networks. Cover collapse of Cypresshead sandy clays gradually expanded small lakes into large lake systems with some lakes joining together to form regional systems. Drainage within the interior parts of the county is internal via karstic systems. Tributary creeks appear along the St. Johns River floodplain along the eastern boundary.

Hernando County exposes the Middle Eocene Ocala Limestone and Oligocene Suwannee Limestone along the Gulf coastal shelf where sea tidal cycles and storm surges eroded most of the Quaternary undifferentiated beach sands. Estuaries line the Gulf Coast, forming on top of limestone.

Paralleling the coastal carbonate platform lies a linear ridge composed of sand dune deposits where the underlying limestone collapsed the sands into shallow sinkholes in the central west part of the dune deposit. The open water lake at the southwestern corner belongs to fresh water wetlands which formed on top of the Weeki Wachee Prairie. In the central part of the county, Suwannee Limestone was exposed when the Hawthorn Group was eroded from the north central part of the region. Erosion occurred after the former Tampa Bay receded following the Miocene Epoch leaving behind remnant ridges supported by Hawthorn Group clay. A large exposure of Hawthorn Group sandy clay and clay remains exposed in the south central part of the county. Some sinkhole lakes formed after sea levels withdrew into the Gulf at the end of the Pleistocene.

In the east central part of the county, Tertiary undifferentiated sand covers the Hawthorn and Suwannee Limestone formations. These sands were left behind after sea levels retreated back to the west during the Quaternary Period. Much of the western part of the county was covered by the Gulf at the time these sands were deposited. Sinkhole caves and caverns collapsed the Hawthorn Group clay and Tertiary sand into sinkholes which produced steeply rolling hills visible along US 301. In the eastern part of the county, Ocala Group limestone is exposed at the surface, pot marked by high concentrations of shallow surface type sinkholes penetrating the upper parts of the limestone unit. Parts of the limestone are covered by remnants of Quaternary dune sands left behind from former beach fronts.

Polk County geology exposes Miocene and younger aged rocks throughout most of the county with the exception of a limited exposure of Upper Eocene Ocala Group limestone and Oligocene Suwannee Limestone in the extreme northwestern corner. Miocene aged Hawthorn Group Bone Valley Member clays contain a vast amount of fossil bones and shells at the time the Miocene Tampa Bay covered the western part of the county. Pliocene aged Cypresshead Formation sandy clays form narrow ridges on top of the Hawthorn Peace River Formation Bone Valley Member in the central western and central parts of the county. Flatlands, occurring between the ridges, represent weathered, eroded, transported, and deposited (aka reworked) Cypresshead Formation sediments along the western flanks of the Lake Wales Ridge. The Lake Wales Ridge is supported by Cypresshead Formation sandy clays along the eastern spine of the county, east of US 27.

Erosion transported some Cypresshead sediments east of the ridge which were covered over by later Tertiary dune sands and Quaternary undifferentiated sands along the eastern parts of the county. In the southwestern part of the county, the open water bodies shown in blue represents phosphate mining from the Bone Valley Member. In the northern part of the western, central, and eastern regions, sinkhole depressions formed large lake systems from multiple collapsing of the older Ocala and Suwannee limestones. The overlying Cypresshead sediments collapsed into caves, caverns, and fracture systems. A narrow sand bar east of Lake Arbuckle in southeast Polk County formed the present divide between the Kissimmee River and Lake Istokpoga drainage.

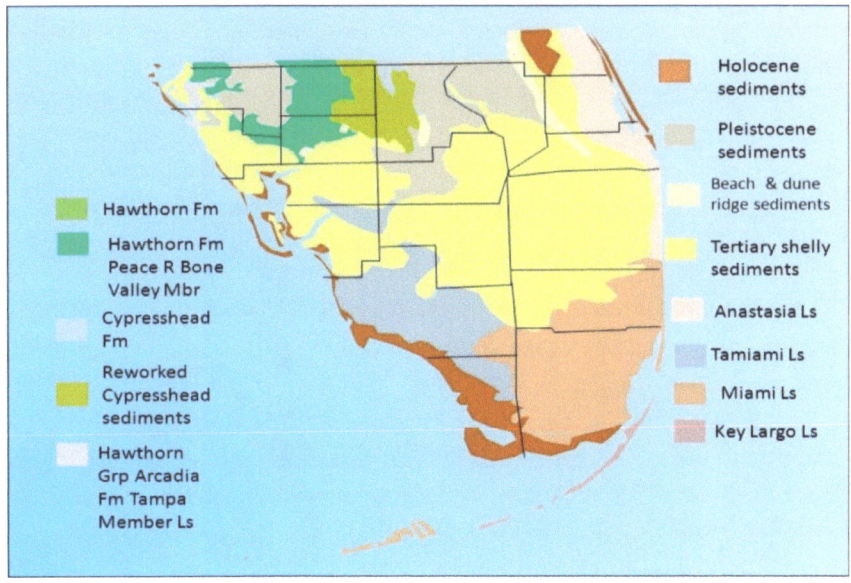

The South Florida Embayment played a major role in focusing sedimentary deposition from the Lake Okeechobee region southward into the Everglades region. During the Pliocene Epoch, the Gulf Coast region accumulated silicicastic sediments mixed with carbonate deposition, particularly from Charlotte Harbor southward into the Florida Keys. Sedimentation began to accumulate in the Okeechobee Basin region. South of Lake Okeechobee, a large lagoon occupied the landscape edged by barrier islands and reefs. Warm, shallow marine waters were protected from storm surges and rough wave activity.

When Pleistocene sea levels covered the region, silicilastic and carbonate deposition expanded eastward into central and eastern Florida from Lake Okeechobee southward to the edges of the Miami Dade County area. Carbonate deposition dominated in the Miami area with mixtures of silicilastic deposition. Miami and Key Largo Limestone accumulated during the Late Pleistocene Epoch.

South of Charlotte County, the **Silver Bluff** terrace re-emerges from its northern pinch out in Citrus County, widening considerably beneath the southern Everglades region. Pliocene to Pleistocene aged sand marls and muds were left behind when seas retreated from the state's interior. The Silver Bluff terrace occupies elevations between 1 and 10 feet above mean sea level.

Coastal Lowlands continue into Sarasota County, widening again through the northern Big Cypress-Everglades region of South Florida. The Big Cypress Swamp is underlain by the **Talbot** terrace surrounded by Everglades Pamlico terrace sands.

Talbot terrace sediments extend into the southwestern central lowlands from Hernando southward into Lee and Glades Counties. The Corkscrew Swamp region of South Florida and the Okeechobee Plain connect the terrace to the Atlantic Coastal Lowlands along the eastern coast.

The **Pamlico** terrrace widens beneath the Big Cypress Swamp and the Northern Everglades regions. Pleistocene fresh and salt water marsh silts and marls (shelly sands) cover older limestone reef deposits that lie very close to the surface. Within South Florida, limestones appear at the surface covered over by a few feet of water during the summer wet season. Limestones are very often hidden by salt marshes, mud flats, marsh grasses, and mangroves.

Manatee County geology consists of Miocene aged rocks belonging to the Hawthorn Group Arcadia and Peace River Formations. Siliciclastic sediments began mixing in with carbonates forming in a shallow restricted embayment setting at this time. At the time of Arcadia Formation sedimentation in the Early Miocene Epoch, siliciclastic sedimentary deposition remained stable covering the entire county. By Late Miocene time, Peace River Formation deposition was completed. Erosion, transport, and deposition redistributed phosphate deposits throughout the county. Most phosphate deposition was concentrated in the eastern and southeastern parts of the county.

When Quaternary sea levels receded from the eastern part of the county, undifferentiated sands were left behind covering the Late Miocene Peace River Formation. Drainage networks formed tributary streams feeding into trunk river systems which carried sediments westward towards the retreating Gulf of Mexico. Along major river channels, and floodplains, Quaternary undifferentiated shell beds accumulated in the western part of the county.

Shifting sea levels back and forth across the western county Miocene platform eroded and redistributed sands, exposing Miocene aged carbonates in the process. Beach and dune sands formed the barrier island system along the western coastline, allowing marsh type estuaries to develop along the inner intracoastal waterway lagoon.

Osceola County surface geology represents conditions occurring when Pleistocene seas withdrew from this part of South Florida. In the northwest corner, Cypresshead Formation sandy clays were deposited as part of the Lake Wales Ridge system during the Pliocene Epoch when seas covered most of Osceola County. The ridge was a peninsula at the time when seas covered flatlands in the remaining central and eastern area of the county. When sea levels retreated to the east, successively lower shoreline terraces were formed by scarps and ridges from former beach face dunes. The Sunderland, Wicomico, Penholoway, Talbot, and Pamlico shorelines developed from west to east across the county.

Pliocene and younger aged sediments covered over the lower Miocene and Eocene clays and limestone. The Ocala Limestone was eroded to a thin layer in the northwest and north central part of the county. The Hawthorn Group overlies the Ocala Limestone and undifferentiated Pleistocene and Holocene sediments cover the Ocala Limestone unit.

In the southern part of the county, thin Pliocene Tamiami Limestone may appear on top of the Hawthorn Group. Much of the Hawthorn Group was eroded during Pleistocene sea level stands in the northwestern part of the county, affecting the distribution of current karst activity in the western parts of the county.

In the eastern parts of the county, flatlands are occupied by former beach dune ridges and linear trending lagoons and estuary type meandering rivers which currently appear as swamps. Beach dune ridges have been eroded and redistributed into flatlands after seas levels retreated from the landscape.

Drainage patterns in the central and eastern parts of the county appear chaotic due to the artifical straightening out of natural creek and river tributaries by canal systems. Much of the open areas of the county belong to agricultural grazing land, altered to drain the flatlands.

The *Osceola Plain* is underlain by the Penholoway terrace. Remnants of the higher Wicomico terrace lie near the center of Osceola County, in the northeast corner of Okeechobee County, and in the northwest region of Highlands County. The Talbot terrace lies along the eastern edge of the plain.

Martin County geology represents conditions present during sea withdrawal to the east. Lake Okeechobee was part of a large expansive lagoon covering the Everglades region of South Florida. Withdrawing seas left behind sheets of shelly sand deposits representing former shallow water offshore and lagoonal facies with linear sand dune deposits positioned in the Attapullah Flats region of the county. Back beach lagoons also developed behind beach dune deposits.

Seas continued to retreat eastward. Anastasia Limestone formed when barrier reefs formed in shallow warm seas rimming the Everglades lagoon during the retreat of late Pleistocene sea levels when seas approached the Atlantic Ocean coastline. Many of the old back beach lagoons became trapped inland, marking the position of the retreating shoreline between the late Pleisotocene and Holocene Epochs. These lagoons became wetland strands resting on top of the eastern part of the Anastasia Limestone.

During Holocene time, seas began to stabilize along the present day coastline, forming barrier beach islands and sand dunes along the coastline. Many of the islands formed present day lagoons marked by the present day Indian River Lagoon and intracoastal waterway. Drainage is mostly controlled by artifical canals which feed the Okeechobee Waterway Canal paralleling FL 76. The canal eventually discharges into Lake Okeechobee. Artificial canals are also present in the south central part of the county, serving agricultural purposes.

Highlands County geology tells a story of late sea withdrawals during the end of the Pleistocene Epoch. The Lake Wales Ridge system extended into the northwestern part of the county, remaining above sea level as a peninsular island. When sea level began to recede to the east, west, and south, erosion transported and deposited Pliocene Cypresshead sediments along the flanks of the ridge. The Miocene Hawthorn Group Peace River Formation peaks out in the extreme southwestern corner, formerly covered over by the reworked Cypresshead sediments. Sheet sands were deposited at the end of the Tertiary into the Quaternary period during sea retreat. Ancient beach dunes accumulated down the central spine along the eastern edge of the more resistant Cypresshead sediments.

Remnant lakes dried up during the Holocene when sea level reached present day positions leaving behind silts and clays in the southwest and north central region. With retreat of the sea to the Atlantic Ocean and Gulf of Mexico, sinkholes began to form when the Peace River Formation limestones began collapsing the overlying Cypresshead sediments along the eastern flanks of the Lake Wales Ridge.

The large lake in the center part of the county is Lake Isotokpoga, an enlarged sinkhole which collapsed during many subsidence cycles. The county is largely agricultural with many artificial canals systems constructed to interrupt and diverted natural creek flow patterns. The Kissimmee River marks the eastern boundary of the county which has been altered in the past but is actively being restored to its original channel.

Hardee County geology consists of Middle Miocene Peace River limestone and Peace River Formation Bone Valley Member clays in the northwestern and southwestern quadrants within the county. When sea levels declined following the Miocene Epoch, many rivers and tributaries developed which began the erosion and transport of the Lake Wales Ridge Cypresshead Formation sandy clays to the west during the Pliocene Epoch. Reworked Cypresshead sediments were deposited on top of Peace River Formation carbonates and siliciastics which were later eroded by the Peace River system in the western part of the county. Quaternary sands accumulated on top of Miocene and Pliocene formations which were also removed by erosion occurring from the Peace River trunk and tributary system. The Peace River continues to cut into the limestone. Most of the landscape supported by Peace River Formation sedimentary deposits exhibits solution type sinkholes occurring in the upper surface of limestone.

Sinkhole activity tends to decrease where clay deposits are thin to absent. Marine shoreline terraces are exposed along the river and tirbutary floodplains excavated by erosion from river downcutting. The Penholoway terrace occupies Peace River system floodplains followed by the Wicomico terrace occupying the headwater regions in the central parts of the county. The Sunderland terrace is supported by reworked Cypresshead sediments in the central northern region and by the Peace River Bone Valley Member clays in the northwestern region. Patches of Quaternary undifferentiated sands make up rolling hills on top of the Peace River Formation.

DeSoto County geology exposes the Miocene Peace River Formation limestone in the northwestern quadrant of the county, and a small localized area in the southeastern part of the county. When seas retreated at the end of the Miocene, rivers drained the southern part of the Lake Wales Ridge. Cypresshead sediments were eroded, transported, and redeposited in the northeastern part of the county following the Pliocene sea retreat. Pleistocene seas returned covering the county again, depositing Quaternary shell beds across the region.

When these seas retreated for the last time, Quaternary undifferentiated sands were left behind when former beach deposits covered the county.

Sea retreat brought about a change in river discharge. The Peace River accelerated erosion by cutting downward into Quaternary deposits until it reached more of the resistant Peace River limestone unit. Erosion, and transport removed much of the Quaternary sands, shell beds, and post-Pliocene Cypresshead reworked sediments exposing the Peace River limestone within many trunk and tributary river channels and floodplains.

Hendry County remained submerged during the Pleistocene, one of the earlier regions to emerge from withdrawing seas. Most of the county was covered over by Tertiary aged sands with a small region in the central eastern part covered by Quaternary aged sands left behind when seas withdrew towards the east. A small exposure of the older Pliocene aged Tamiami Limestone became exposed by the downcutting Caloosahatchee River west of La Belle. Drainage is mostly internal, occurring by canals constructed to lower ground water for agricultural purposes.

Collier County geology exposes a large area of Pliocene Tamiami Limestone in the central north and southwest section of the county. Quaternary shell beds cover the limestone in the northwest, northeast, and eastern parts of the county. The county remained submerged while the northern parts of South Florida began to dry out. Sandy limestones containing sand beds mixed with clay and marl were deposited in shallowing sea levels subjected to open circulation which allowed phosphate to mix in with carbonate and siliclastic mixtures.

When sea levels began to retreat towards the Pleistocene Okeechobee Lagoon to the east, shell beds were deposited in the eastern part of the county. Portions of south and eastern Collier County remained flooded by shallow water covering the Big Corkscrew Swamp, and Big Cypress Swamp sections of the Everglades. Coastal mangrove swamps developed along the quiet, low energy coastline of the Gulf of Mexico during Holocene time.

Miami Dade geology is dominated by the Everglades which covered the entire county after the Pleistocene sea withdrawal. The last sea retreat covered the county at the end of the Pleistocene Epoch. The City of Miami southward to Homestead, was built on top of fill placed on Miami Limestone. The Everglades covers the region of the county west of FL 991 south to Florida Bay.

Quaternary shell beds occupy the extreme northwest corner marking the southeastward extent of the Lake Okeechobee lagoonal plain.

The Everglades were fed by flooding of the lake, flowing as a sheet on top of the Miami Limestone towards the southeast curving to the southwest direction, marked by the hammock island orientations.

When seas withdrew to the east following the Pleistocene, reefs began to form offshore along the southeastern edge of the Florida peninsula. Complete withdrawal of the seas to current elevations left these reefs exposed to aerial erosion. The Key Largo Limestone formed under these conditions, supporting the eastern key reef system including Key Largo and the upper key islands. Holocene mangrove swamps trapped silts and clays along the northern edge of Florida Bay, protecting the Florida Key reefs from siltation and tropical storm erosion. The lower keys are supported by Miami Limestone.

Part II. Geologic Maps

Each of the county geologic maps were compiled using the Florida Geological Survey's Open File Map Series Nos. 3 through 68 as a base. Each lithologic unit was color coded to match the units presented on the maps. The legends on each map were described as stated in the explanations contained on the map with a description expanding on the unit included below the map. The geologic legend symbol was included in the parentheses after the designated unit name.

The township and range designation were added without sections to clear the map clutter which tends to interfere with identifying lithologic units on the state maps. The township and range sections follow the accepted principle designates of 1 thorugh 36 with section 1 starting in the upper right corner and section 36 ending in the lower right corner. On occasion, certain maps have distorted configurations of township and range due to map preparation distortions associated with enlarging the map scale for the purpose of presenting a larger image of the lithologic units within each county. Examples are provided in **Figure 7**.

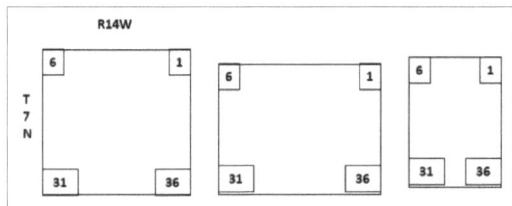

Figure 7. Examples of township and range geometrical configurations indicating map distortion direction during preparation. The left shape represents a normal configuration. The middle represents horizontal distortion. The right represents vertical distortion.

The State of Florida was subdivided into three divisions: The Panhandle, North Peninsular, and South Peninsular sections. Each map presented was arranged in alphabetical order for each division. Geologic maps represent surfical geology at or near the surface. Monroe County was separated into two maps: the first map is for the mainland portion and Key Largo while the second map pertains to the Florida Keys.

Panhandle Florida Maps

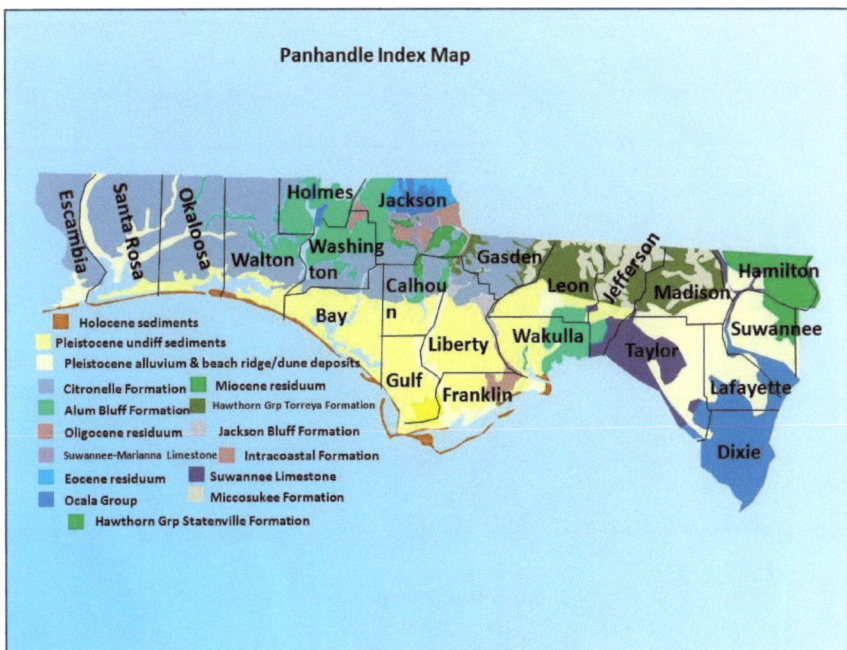

Figure 8. Panhandle Florida Index map presenting the counties included in this section. The eastern cutoff was selected randomly at the Suwannee River except for Suwannee County which was included as part of the Panhandle counties. Twenty two counties make up the district.

Bay County

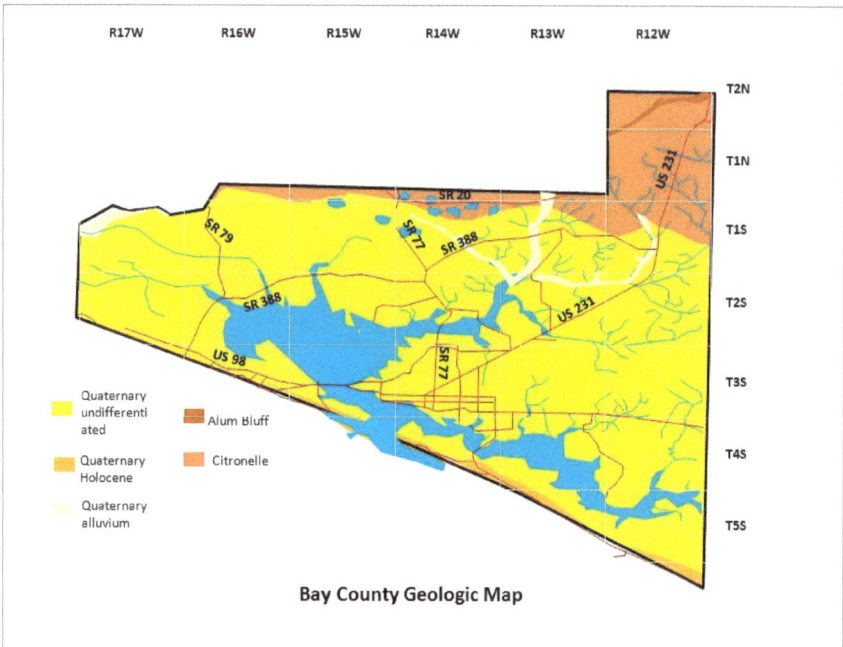

Figure 9. Bay County surfical geologic map. Source: Revised from Florida Geological Survey Open File Map No. 19.

Quaternary

Quaternary undifferentiated sand (Qu)- Undifferentiated quartz sands which may consist in part of reworked Citronelle Formation fine to coarse grained with varying percentages of silt and clay.

Quaternary Holocene (Qh)- Holocene sediments consisting of quartz sand with minor amounts of organic matter and clay associated with lagoonal deposits consisting of mostly beach and dune sands along present day coastline.

Quaternary alluvium (Qal)- Alluvium deposited in river flood plains. Quartz sands, silts, and clays along with varying percentages of organic matter are present. Often consists of reworked Citronelle Formation and older units very fine to coarse grained.

Citronelle Formation (QTci)- Fine to coarse grained sands with gravel, silt, and clay often oxidized to reddish hues in exposures.

Tertiary

Alum Bluff Group (Tab)- Undifferentiated Alum Bluff Group consisting of clays, silts, and sands often greenish to gray hues. May contain shells.

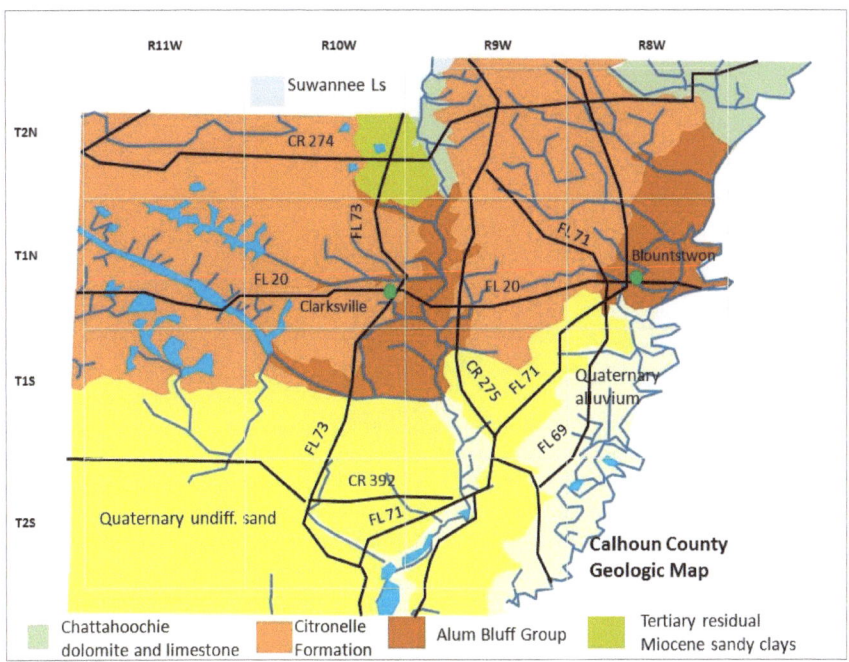

Figure 10. Calhoun County geologic map. Revised from Florida Geological Survey Open File Map No. 14.

Quaternary

Quaternary undifferentiated sand (Qu)- Undifferentiated quartz sands which may consist of reworked Citronelle Formation. Consists of fine to coarse grained with varying percentages of silt and clay.

Tertiary

Citronelle Formation (Tci)- Fine to coarse grained sands with gravel, silt, and clay often oxidized to reddish hues in exposures. Reworked phase of this unit is designated QTci.

Alum Bluff Group (Tab)- Undifferentiated Alum Bluff Group consisting of clays, silts, and sands often greenish to gray hues. May contain shells.

Chattahoochie Dolomite and Limestone (Tchat)- Dolostones with subordinate limestone, clays, and silts often fossiliferous and moldic.

Tertiary residual sandy clays (Trm)- Residuum on Miocene consisting of clay, sandy clay, and clayey sand resulting from the dissolution of Miocene carbonates and lowering/reworking of Tertiary siliciclastic sediments.

Dixie County

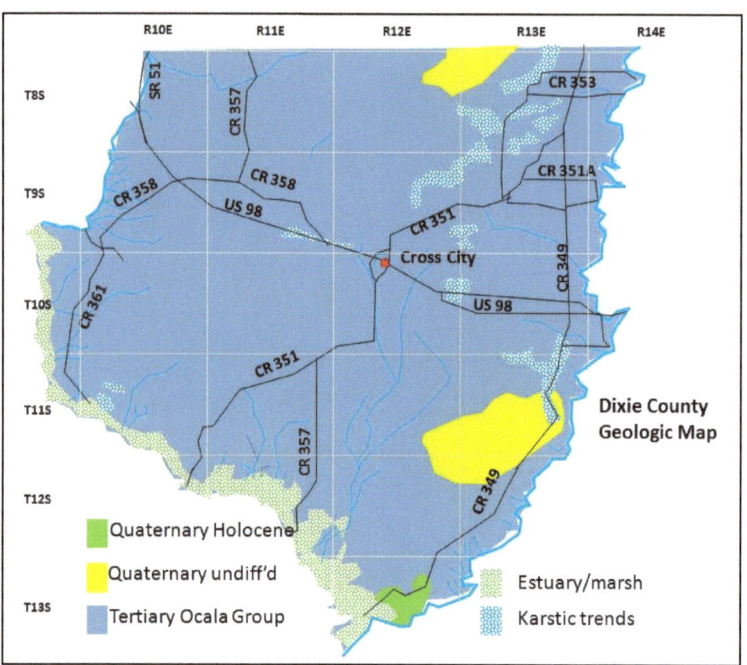

Figure 11. Dixie County geologic map. Revised from Florida Geological Survey Open File Map No. 35.

Quaternary

Quaternary undifferentiated sand (Qu)- Undifferentiated surficial sands, clayey sands, clays, marls, and peats greater than 20 feet thick. No formations are recognized.

Quaternary Holocene (Qph)- Undifferentiated Holocene sediments includes beach, lagoonal, and marsh sediments. Similar sediments may be designated as Qh.

Estuary/marsh- deposits occurring along Gulf of Mexico coastal margins.

Karstic trends- areas where Ocala Group limestone are at or near the surface exhibiting collapse features reaching the surface.

Tertiary

Ocala Group Limestone (To)- White to gray, fossiliferous, moldic limestone. Varies from packstone to grainstone.

Escambia County

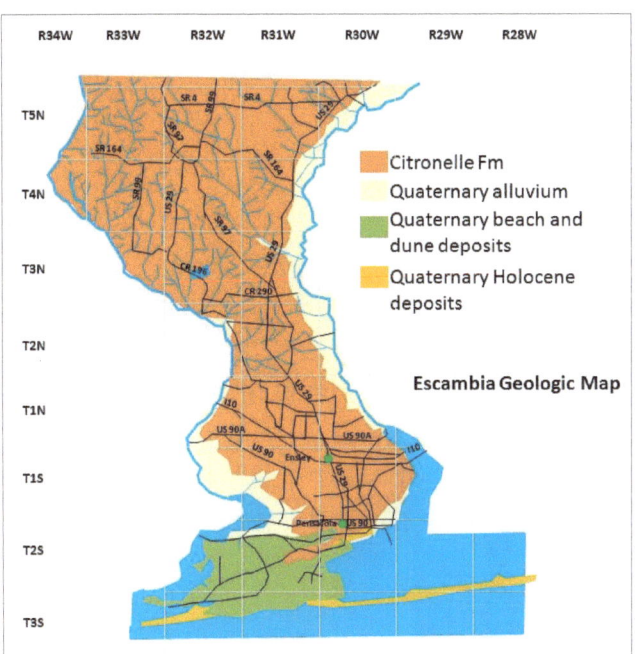

Figure 12. Escambia County geologic map. Revised from Florida Geological Survey Open File Map No. 14.

Quaternary

Quaternary alluvium (Qal)- Alluvium deposited in the river flood plains consisting of quartz sands, silts, and clays with varying percentages of organic matter consisting of reworked Citronelle Formation and older units. Very fine to coarse grained.

Quaternary beach & dune deposits (Qbd)- Quartz sands with surface expressions of beach ridges and dunes. A geomorphic unit resting on undifferentiated, often clean quartz sands. Generally fine to medium grained.

Quaternary Holocene (Qh)- Holocene sediments consist of quartz sand with minor amounts of organic matter and clay associated with lagoonal deposits. Mostly, these sediments consist of beach and dune sands along the present coastline.

Quaternary/Tertiary

Quaternary/Tertiary Citronelle Formation (QTci)- Fine to coarse grained sands with gravel, silt, and clay. Often oxidized to reddish hues in exposures.

Franklin County

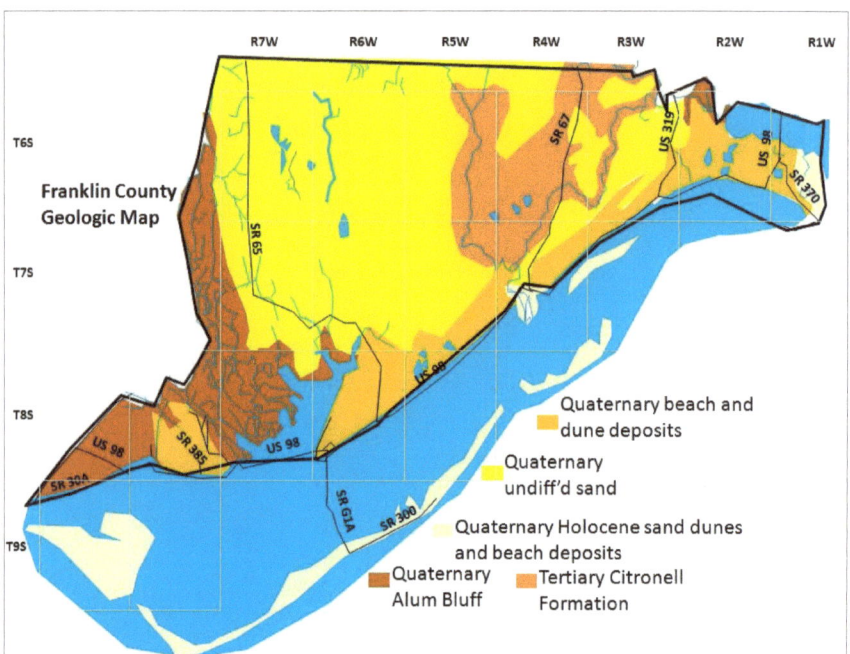

Figure 13. Franklin County geologic map. Revised from Florida Geological Survey Open File Map No. 21.

Quaternary

Quaternary undifferentiated sand (Qu)- Undifferentiated quartz sands possibly consisting in part of reworked Citronelle Formation. Fine to coarse grained with varying percentages of silt and clay.

Quaternary beach & dune deposits (Qbd)- Quartz sands with surface expressions of beach ridges and dunes. A geomorphic unit on undifferentiated, often clean quartz sands. Generally fine to medium grained.

Quaternary Holocene deposits (Qh)- Holocene sediments consisting of quartz sand with minor amounts of organic matter and clay associated with lagoonal deposits. Mostly consists of beach and dune sands along present coastlines.

Tertiary

Tertiary Alum Bluff (Tab)- correction made to the map which lists the Alum Bluff as Quaternary. The FGS map lists the Alum Bluff Group as Tertiary in Franklin County which is undifferentiated clays, silts, and sands often greenish in gray hues. May contain shells.

Tertiary Citronelle Formation (Tci)- Fine to coarse grained sands with gravel, silt, and clay. Often oxidized to reddish hues in exposures.

Gadsden County

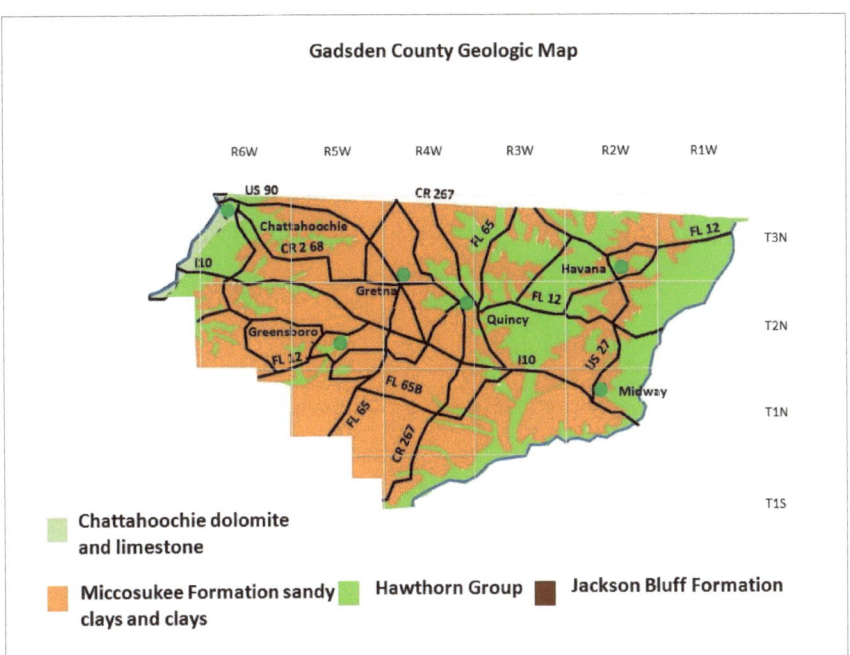

Figure 14. Gadsden County geologic map. Revised from Florida Geological Survey Open File Map No. 22.

Tertiary

Chattahoochie Dolomite and Limestone (Tchat)- Dolostones with subordinate limestone, clays, and silts that are fossiliferous with abundant mold casts.

Hawthorn Group (Th)- Sandy clays, silt with occasional carbonates. May contain phosphate locally.

Jackson Bluff Formation (Tjb)- Variably sandy clay often with abundant fossils, greenish gray to light olive green.

Miccosukee Formation (Tmc)- Clayey sand to sandy clay with clay stringers and lenses, occasionally cross bedded, reddish orange to brick red, oxidized, poorly to moderately consolidated.

Gulf County

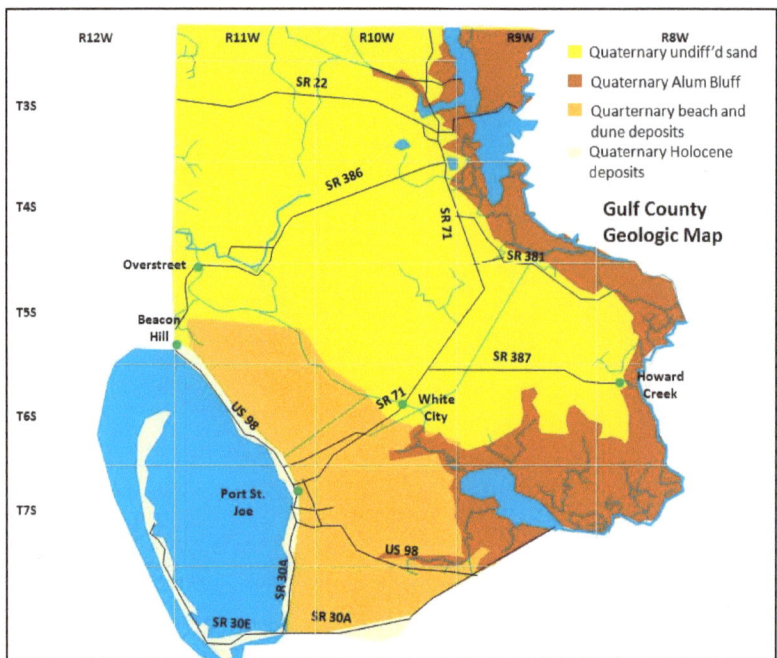

Figure 15. Gulf County geologic map. Revised from Florida Geological Survey Open File Map No. 23.

Quaternary

Quaternary undifferentiated (Qu)- Undifferentiated quartz sands possibly consisting in part of reworked Citronelle Formation. Fine to coarse grained with varying percentages of silt and clay.

Quaternary beach & dune deposits (Qbd)- Quartz sands with surface expressions of beach ridges and dunes. A geomorphic unit on undifferentiated, often clean quartz sands. Generally fine to medium grained.

Quaternary Holocene (Qh)- Holocene sediments consisting of quartz sand with minor amounts of organic matter and clay associated with lagoonal deposits. Mostly consists of beach and dune sands along present coastlines.

Tertiary

Tertiary Alum Bluff (Tab)- correction made to the map which lists the Alum Bluff as Quaternary. The FGS map lists the Alum Bluff Group as Tertiary in Gulf County which is undifferentiated clays, silts, and sands often greenish in gray hues. May contain shells.

Hamilton County

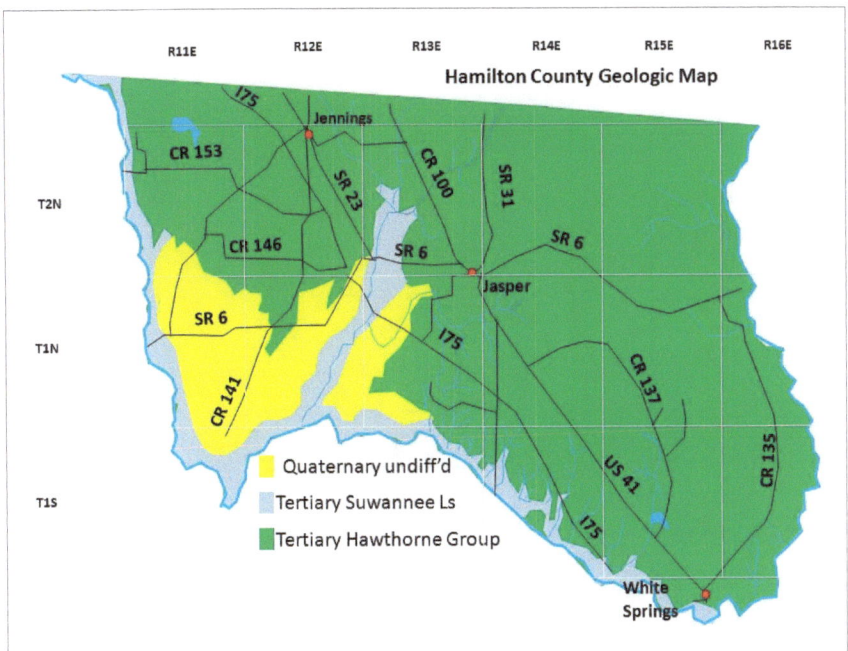

Figure 16. Hamilton County geologic map. Revised from Florida Geological Survey Open File Map No. 32.

Quaternary

Quaternary undifferentiated (Qu)- undifferentiated surficial sands, clayey sands, clays, marls, and peats greater than 20 feet thick. No formations recognized.

Tertiary

Tertiary Hawthorn Group (Th)- Sandy clays, silt, ocassional carbonates which may contain phosphate locally.

Tertiary Suwannee Limestone (Ts)- Marine limestone, fossiliferous, varies from poorly recrystallized to well recrystallized packstone to grainstone with minor siliciclastics present.

Holmes County

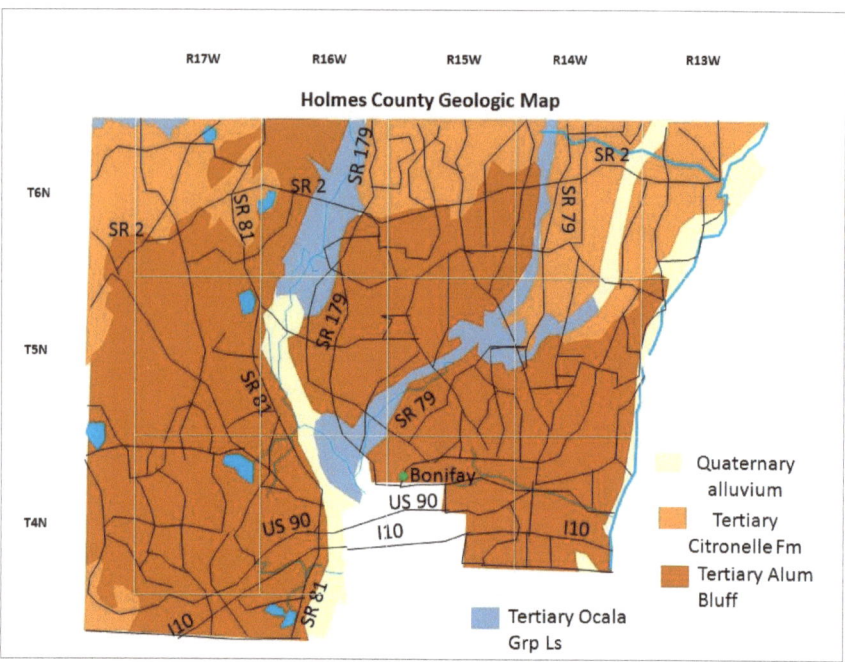

Figure 17. Holmes County geologic map. Revised from Florida Geological Survey Open File Map No. 24.

Quaternary

Quaternary alluvium (Qal)- Alluvium deposited in the river flood plains. Quartz sands, silts, and clays with varying percentages of organic matter. Often consists of reworked Citronelle Formation and older units. Very fine to coarse grained.

Tertiary

Tertiary Alum Bluff (Tab)- Undifferentiated Alum Bluff Group consisting of clays, silts, and sands often greenish to gray hues. May contain shells.

Tertiary Citronelle Formation (Tci)- Fine to coarse grained sands with gravel, silt, and clay. Oftne oxidized to reddish hues in exposures. The reworked phase of this is designated as QTci.

Tertiary Ocala Group (To)- Ocala Limestone white to gray, fossiliferous molds, varying from packstone to grainstone.

Jackson County

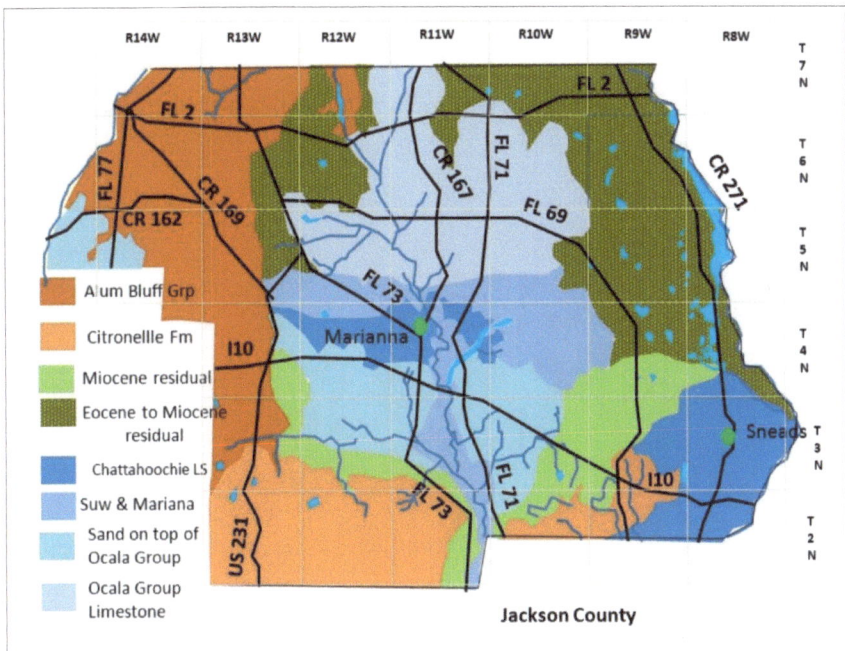

Figure 18. Jackson County geologic map. Revised from Florida Geological Survey Open File Map No. 25.

Tertiary

Tertiary Alum Bluff Group (Tab)- Undifferentiated Alum Bluff Group consisting of clays, silts, and sands often greenish to gray hues. May contain shells.

Tertiary Citronelle Formation (Tci)- Fine to coarse grained sands with gravel, silt, and clay. Oftne oxidized to reddish hues in exposures. The reworked phase of this is designated as QTci.

Eocene to Miocene residual (Tre)- Clay, sandy clay, and clayey sands resulting from the dissolution of Eocene to Miocene carbonates and lowering/reworking of Tertiary siliciclastic deposits.

Miocene residual (Trm)- Clay, sandy clay, and clayey sand resulting from the dissolution of Miocene carbonates and lowering/reworking of Tertiary siliciclastic deposits.

Tertiary Chattahoochie Limestone (Tchat)- Dolostones with subordinate limestone, clays, and silts that are fossiliferous with abundant mold casts.

Tertiary Suwannee & Marianna Limestone (Tsm)- Undifferentiated Suwannee Limestone and Marianna Limestone. Appears white to pale orange, fossiliferous, molds common, varies from wackstone to grainstone with minor siliciclastics present.

Tertiary Sand on top of Ocala Group – Sand resting on top of Ocala Group Limestone.

Tertiary Ocala Group (To)- White to gray fossiliferous moldic limestone. Varies from packstone to grainstone.

Jefferson County

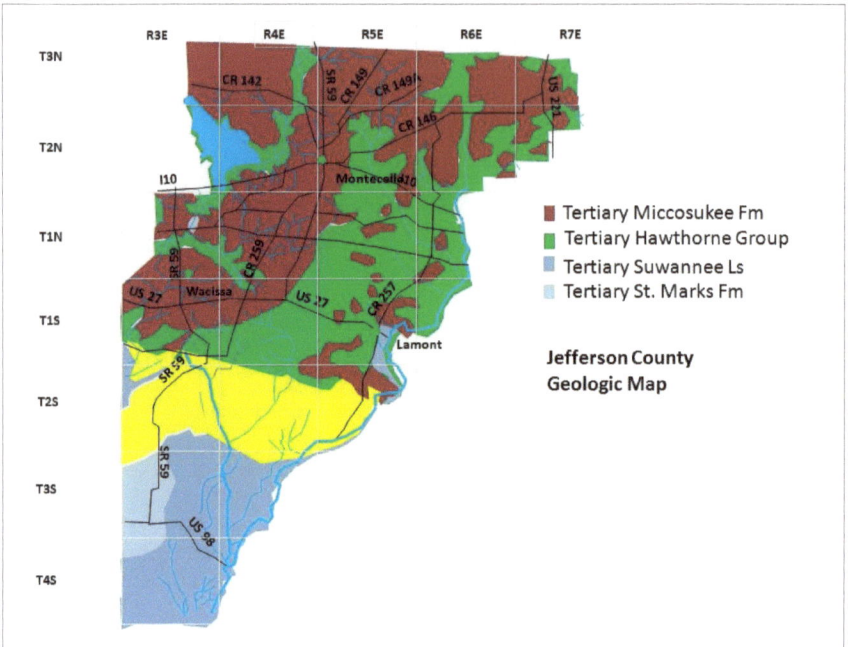

Figure 19. Jefferson County geologic map. Revised from Florida Geological Survey Open File Map No. 31.

Tertiary

Tertiary Miccosukee Formation (Tmc)- Clayey sand to sandy clay with clay stringers and lenses, occasionally cross bedded, reddish orange to brick red, oxidized, poorly to moderately consolidated.

Tertiary Hawthorn Group (Th)- Sandy clays, silt, ocassional carbonates which may contain phosphate locally.

Tertiary Suwannee Formation (Ts)- Marine limestone, fossiliferous, varies from poorly recrystallized to well recrystallized packstone to grainstone with minor siliciclastics present.

Tertiary St. Marks Formation (Tsmk)- Marine limestone, white to tan often sandy, fossiliferous with abundant mollusk molds varying from packstone to wakstone but more muddy than the Suwanne Limestone.

Lafayette County

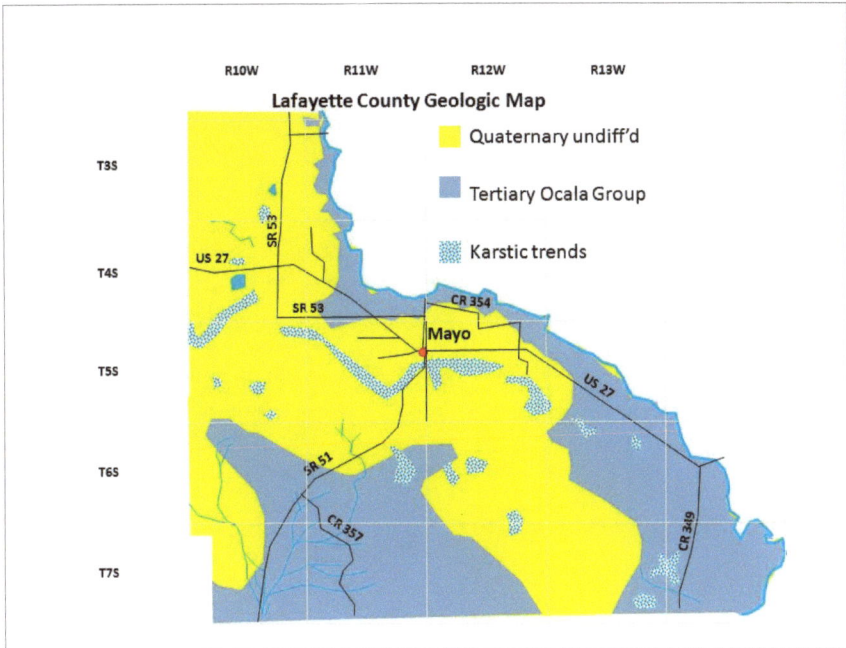

Figure 20. Lafayette County geologic map. Revised from Florida Geological Survey Open File Map No. 34.

Quaternary

Quaternary undifferentiated (Qu)- undifferentiated surficial sands, clayey sands, clays, marls, and peats greater than 20 feet thick. No formations recognized.

Tertiary

Tertiary Ocala Group (To)- White to gray limestone, fossiliferous, moldic, varying from packstone to grainstone.

Leon County

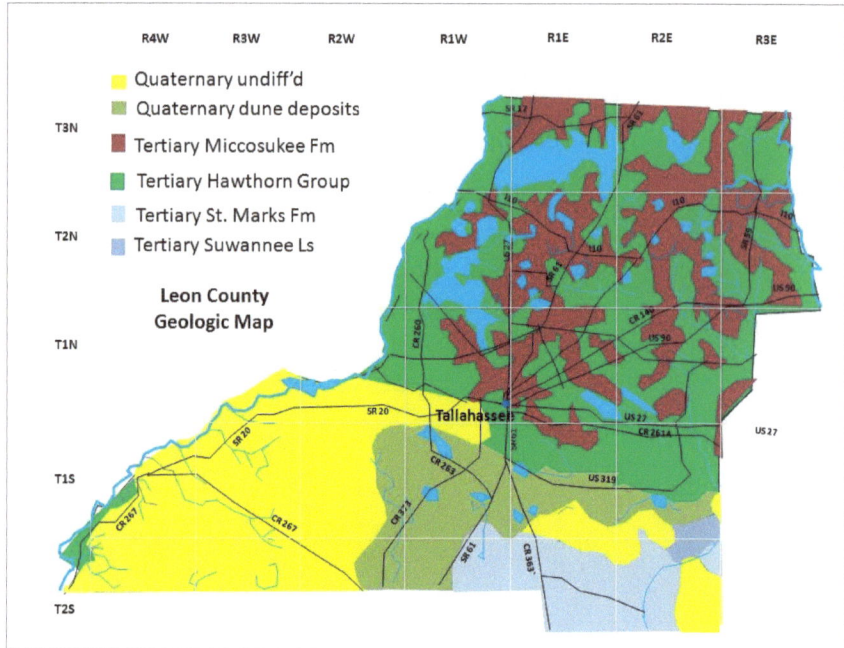

Figure 21. Leon County geologic map. Revised from Florida Geological Survey Open File Map No. 28.

Quaternary

Quaternary undifferentiated (Qu)- Undifferentiated quartz sands. May consist in part of reworked Citronelle Formation. Fine to coarse grained with varying percentages of silt and clay.

Quaternary dune deposits (Qd)- Dune sands mapped where dunes are very well developed.

Tertiary

Tertiary Hawthorn Group (Th)- Sandy clays, silt, ocassional carbonates which may contain phosphate locally.

Tertiary Miccosukee Formation (Tmc)- Clayey sand to sandy clay with clay stringers and lenses, occasionally cross bedded, reddish orange to brick red, oxidized, poorly to moderately consolidated.

Tertiary Suwannee Formation (Ts)- Marine limestone, fossiliferous, varies from poorly recrystallized to well recrystallized packstone to grainstone with minor siliciclastics present.

Tertiary St. Marks Formation (Tsmk)- Marine limestone, white to tan often sandy, fossiliferous with abundant mollusk molds varying from packstone to wackstone but more muddy than the Suwannee Limestone.

Liberty County

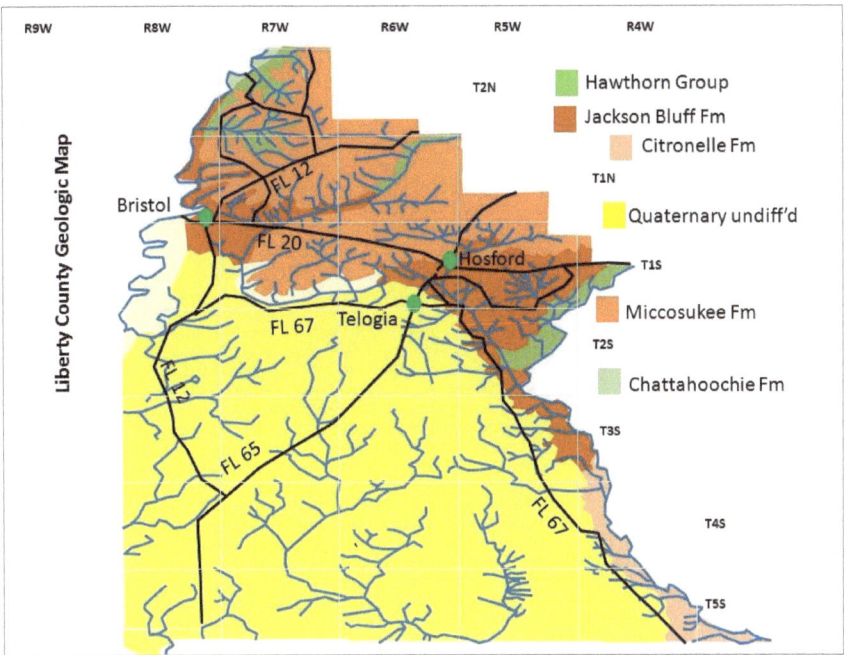

Figure 22. Liberty County geologic map. Revised from Florida Geological Survey Open File Map No. 26.

Quaternary

Quaternary undifferentiated (Qu)- Undifferentiated quartz sands. May consist in part of reworked Citronelle Formation. Fine to coarse grained with varying percentages of silt and clay.

Tertiary

Tertiary Chattahoochie Formation (Tchat)- Dolostones with subordinate limestone, clays, and silts that are fossiliferous with abundant mold casts.

Tertiary Miccosukee Formation (Tmc)- Clayey sand to sandy clay with clay stringers and lenses, occasionally cross bedded, reddish orange to brick red, oxidized, poorly to moderately consolidated.

Tertiary Hawthorn Group (Th)- Sandy clays, silt, ocassional carbonates which may contain phosphate locally.

Tertiary Jackson Bluff Formation (Tjb)- Variably sandy clay often with abundant fossils, greenish gray to light olive green.

Tertiary Citronelle Formation (Tci)- Fine to coarse grained sands with gravel, silt, and clay. Often oxidized to reddish hues in exposures. The reworked phase of this is designated as QTci.

Madison County

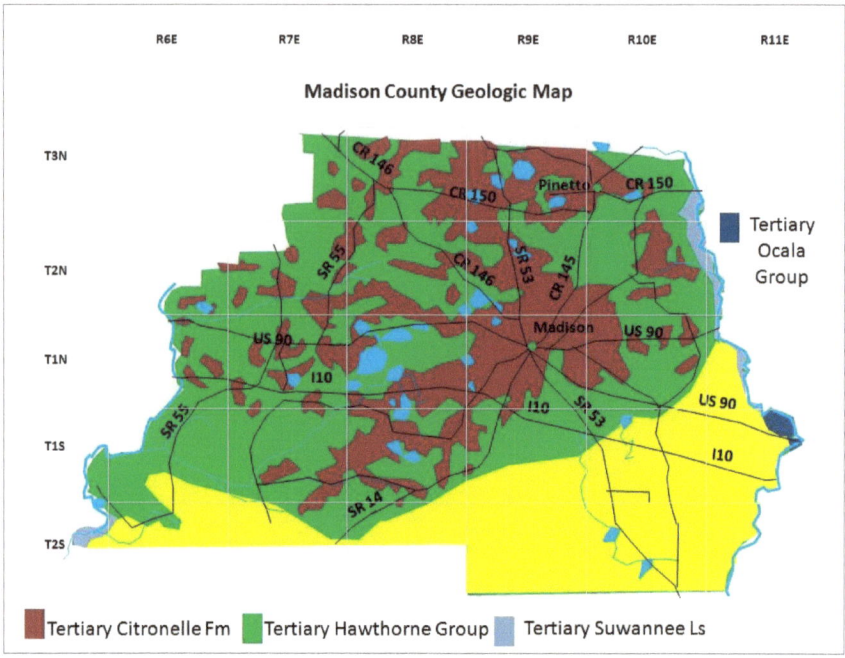

Figure 23. Madison County geologic map. Revised from Florida Geological Survey Open File Map No. 27.

Tertiary

Tertiary Hawthorn Group (Th)- Sandy clays, silt, occassional carbonates which may contain phosphate locally.

Tertiary Citronelle Formation (Tci)- Fine to coarse grained sands with gravel, silt, and clay. Oftne oxidized to reddish hues in exposures. The reworked phase of this is designated as QTci.

Tertiary Suwannee Limestone (Ts)- Marine limestone, fossiliferous, varies from poorly recrystallized to well recrystallized packstone to grainstone with minor siliciclastics present.

Tertiary Ocala Group (To)- White to gray limestone, fossiliferous, moldic, varying from packstone to grainstone.

Okaloosa County

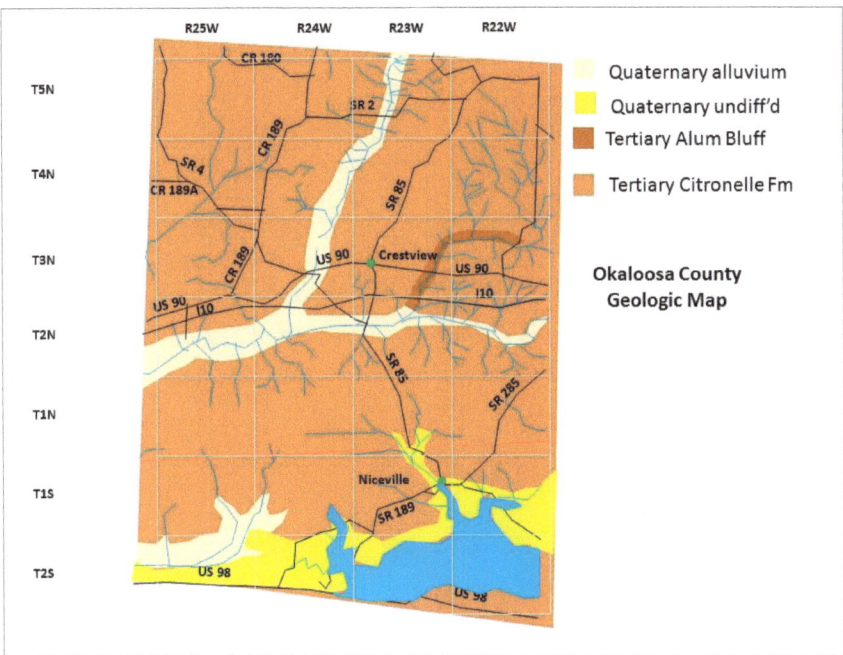

Figure 24. Okaloosa County geologic map. Revised from Florida Geological Survey Open File Map No. 16.

Quaternary

Quaternary undifferentiated (Qu)- Undifferentiated quartz sands. May consist in part of reworked Citronelle Formation. Fine to coarse grained with varying percentages of silt and clay.

Quaternary alluvium (Qal)- Alluvium deposited in the river flood plains. Quartz sands, silts, and clays with varying percentages of organic matter. Often consists of reworked Citronelle Formation and older units. Very fine to coarse grained.

Tertiary

Tertiary Alum Bluff Group (Tab)- Undifferentiated Alum Bluff Group consisting of clays, silts, and sands often greenish to gray hues. May contain shells.

Tertiary Citronelle Formation (Tci)- Fine to coarse grained sands with gravel, silt, and clay. Often oxidized to reddish hues in exposures. The reworked phase of this is designated as QTci.

Santa Rosa County

Figure 25. Santa Rosa County geologic map. Revised from Florida Geological Survey Open File Map No. 15.

Quaternary

Quaternary undifferentiated (Qu)- Undifferentiated quartz sands. May consist in part of reworked Citronelle Formation. Fine to coarse grained with varying percentages of silt and clay.

Quaternary alluvium (Qal)- Alluvium deposited in the river flood plains. Quartz sands, silts, and clays with varying percentages of organic matter. Often consists of reworked Citronelle Formation and older units. Very fine to coarse grained.

Quaternary beach & dune deposits (Qbd)- Quartz sands with surface expressions of beach ridges and dunes. A geomorphic unit on undifferentiated, often clean quartz sands. Generally fine to medium grained.

Tertiary

Tertiary Citronelle Formation (Tci)- Fine to coarse grained sands with gravel, silt, and clay. Often oxidized to reddish hues in exposures. The reworked phase of this is designated as QTci.

Suwannee County

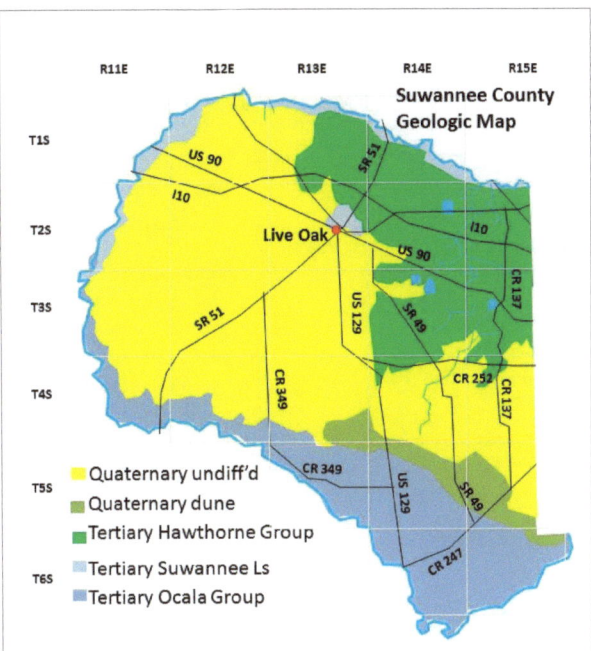

Figure 26. Suwannee County geologic map. Revised from Florida Geological Survey Open File Map No. 33.

Quaternary

Quaternary undifferentiated (Qu)- Undifferentiated quartz sands. May consist in part of reworked Citronelle Formation. Fine to coarse grained with varying percentages of silt and clay.

Quaternary dune deposits (Qd)- Dune sands mapped where dunes are very well developed.

Tertiary

Tertiary Hawthorn Group (Th)- Sandy clays, silt, occassional carbonates which may contain phosphate locally.

Tertiary Suwannee Limestone (Ts)- Marine limestone, fossiliferous, varies from poorly recrystallized to well recrystallized packstone to grainstone with minor siliciclastics present.

Tertiary Ocala Group (To)- White to gray limestone, fossiliferous, moldic, varying from packstone to grainstone.

Taylor County

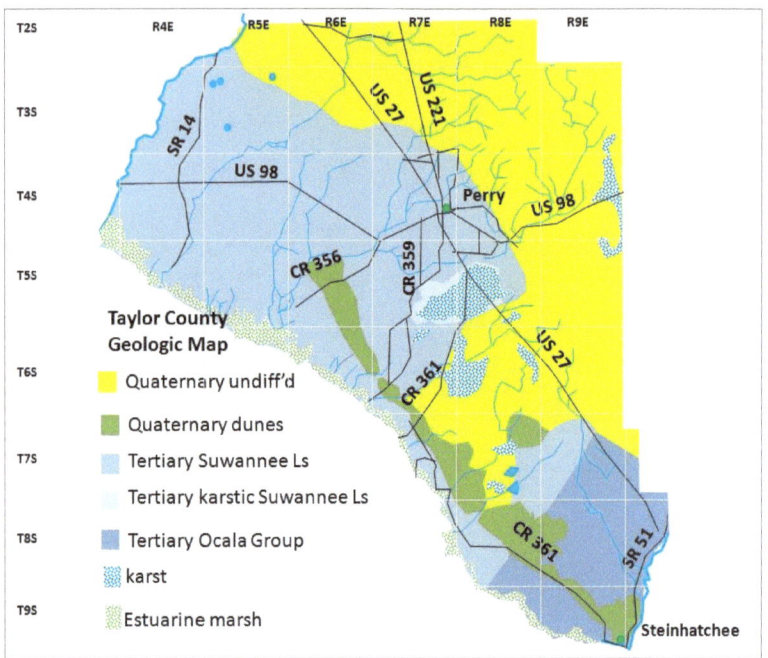

Figure 27. Taylor County geologic map. Revised from Florida Geological Survey Open File Map No. 29.

Quaternary

Quaternary undifferentiated (Qu)- Undifferentiated quartz sands. May consist in part of reworked Citronelle Formation. Fine to coarse grained with varying percentages of silt and clay.

Quaternary dune deposits (Qd)- Dune sands mapped where dunes are very well developed.

Tertiary

Tertiary Suwannee Limestone (Ts)- Marine limestone, fossiliferous, varies from poorly recrystallized to well recrystallized packstone to grainstone with minor siliciclastics present.

Tertiary Karstic Suwannee Limestone (Tks)- Suwannee Limestone, karstified and overlain by variable thicknesses of siliciclastic sediments.

Tertiary Ocala Group (To)- White to gray limestone, fossiliferous, moldic, varying from packstone to grainstone.

Karst – solution sinkholes occurring on the top of the exposed limestone.

Estuarine marsh- salt to brackish marshes occurring along the coastline.

Wakulla County

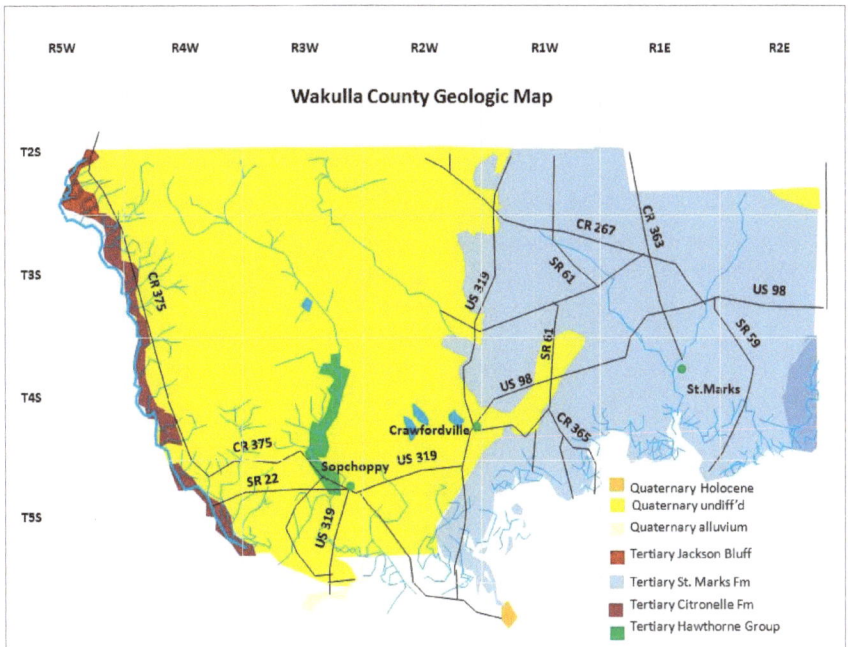

Figure 28. Wakulla County geologic map. Revised from Florida Geological Survey Open File Map No. 30.

Quaternary

Quaternary undifferentiated (Qu)- Undifferentiated quartz sands. May consist in part of reworked Citronelle Formation. Fine to coarse grained with varying percentages of silt and clay.

Quaternary alluvium (Qal)- Alluvium deposited in the river flood plains. Quartz sands, silts, and clays with varying percentages of organic matter. Often consists of reworked Citronelle Formation and older units. Very fine to coarse grained.

Quaternary Holocene (Qh)- Holocene sediments consisting of quartz sand with minor amounts of organic matter and clay associated with lagoonal deposits. Mostly beach and dune sands along present coastline.

Tertiary

Tertiary Jackson Bluff Formation (Tjb)- Variably sandy clay often with abundant fossils, greenish gray to light olive green.

Tertiary Hawthorn Group (Th)- Sandy clays, silt, occasional carbonates which may contain phosphate locally.

Tertiary Citronelle Formation (Tci)- Fine to coarse grained sands with gravel, silt, and clay. Often oxidized to reddish hues in exposures. The reworked phase of this is designated as QTci.

Tertiary St. Marks Formation (Tsmk)- Marine limestone, white to tan often sandy, fossiliferous with abundant mollusk molds, moldic, varying from wackstone to packstone but more muddy than the Suwannee Limestone.

Walton County

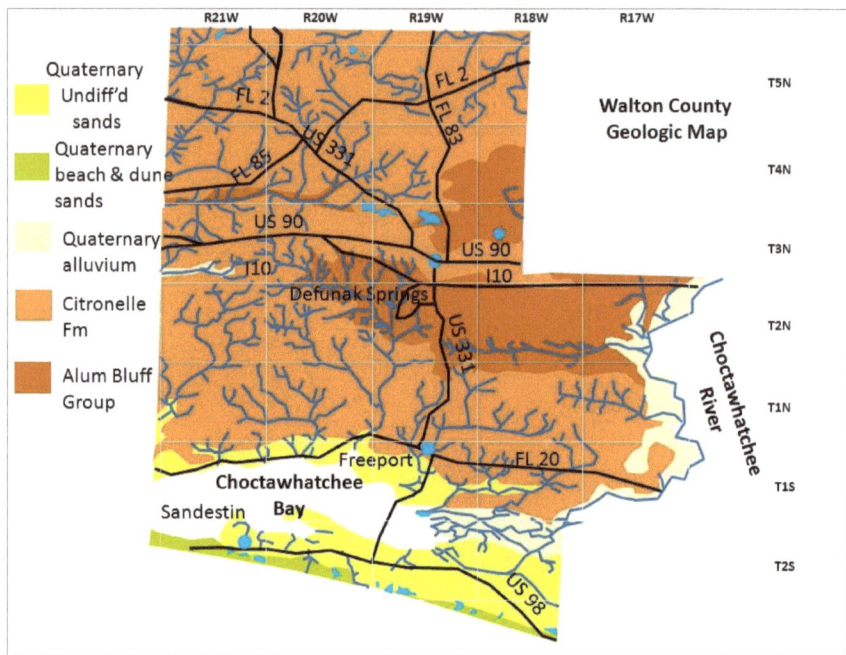

Figure 29. Walton County geologic map. Revised from Florida Geological Survey Open File Map No. 17.

Quaternary

Quaternary undifferentiated (Qu)- Undifferentiated quartz sands. May consist in part of reworked Citronelle Formation. Fine to coarse grained with varying percentages of silt and clay.

Quaternary beach & dune deposits (Qbd)- Quartz sands with surface expressions of beach ridges and dunes. A geomorphic unit on undifferentiated, often clean quartz sands. Generally fine to medium grained.

Quaternary alluvium (Qal)- Alluvium deposited in the river flood plains. Quartz sands, silts, and clays with varying percentages of organic matter. Often consists of reworked Citronelle Formation and older units. Very fine to coarse grained.

Tertiary

Tertiary Alum Bluff Group (Tab)- Undifferentiated Alum Bluff Group consisting of clays, silts, and sands often greenish to gray hues. May contain shells.

Tertiary Citronelle Formation (Tci)- Fine to coarse grained sands with gravel, silt, and clay. Often oxidized to reddish hues in exposures. The reworked phase of this is designated as QTci.

Washington County

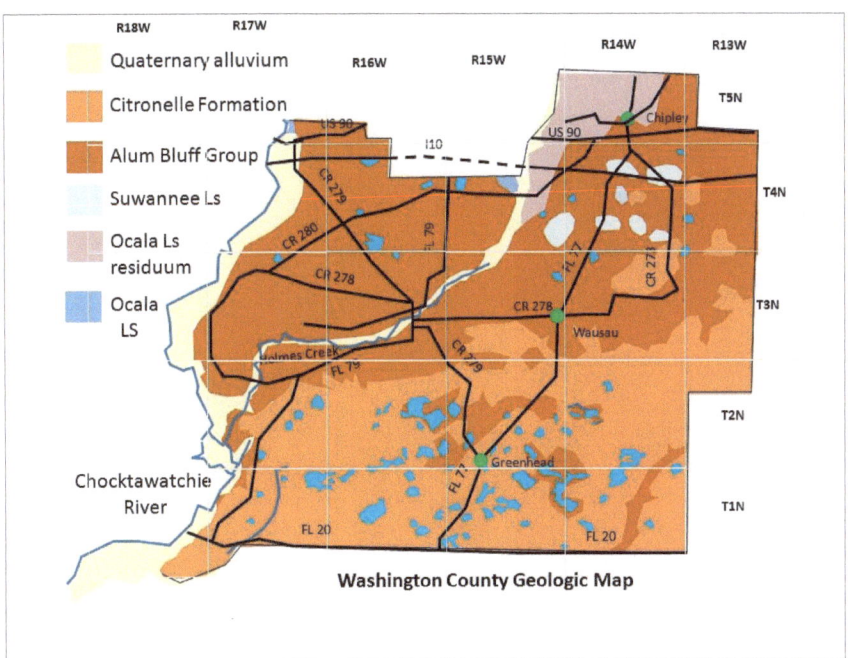

Figure 30. Washington County geologic map. Revised from Florida Geological Survey Open File Map No. 18.

Quaternary

Quaternary alluvium (Qal)- Alluvium deposited in the river flood plains. Quartz sands, silts, and clays with varying percentages of organic matter. Often consists of reworked Citronelle Formation and older units. Very fine to coarse grained.

Tertiary

Tertiary Alum Bluff Group (Tab)- Undifferentiated Alum Bluff Group consisting of clays, silts, and sands often greenish to gray hues. May contain shells.

Tertiary Citronelle Formation (Tci)- Fine to coarse grained sands with gravel, silt, and clay. Often oxidized to reddish hues in exposures. The reworked phase of this is designated as QTci.

Tertiary Suwannee Limestone (Ts)- Marine limestone, fossiliferous, varies from poorly recrystallized to well recrystallized packstone to grainstone with minor siliciclastics present.

Tertiary Ocala Limestone residuum (Tro)- Sandy and clayey residuum of Ocala limestone laying on top of the Ocala Group.

Tertiary Ocala Group (To)- White to gray limestone, fossiliferous, moldic, varying from packstone to grainstone.

Part III. North Peninsular Florida

North Peninsular Florida Index Map

Alachua County

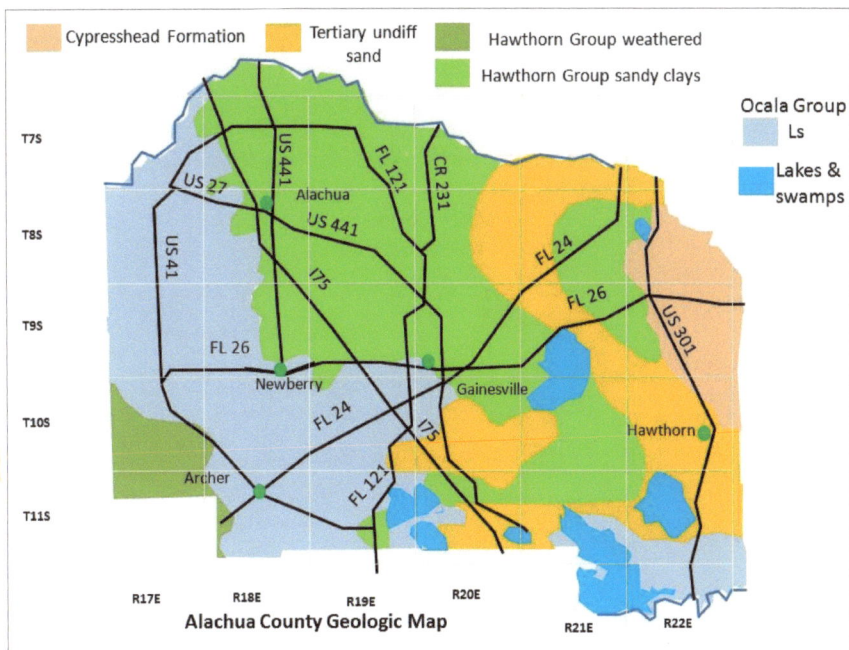

Figure 31. Alachua County geologic map. Revised from Florida Geological Survey Open File Map No. 12.

Tertiary

Tertiary undifferentiated sand (QTu) occurs in the eastern part of the county. These sediments consist of sands with units of silt, clay, and organic matter lying unconformably on the Ocala limestone, Hawthorn Group or Cypresshead Formation. The unit may be partially Quaternary in nature.

Tertiary Late Pliocene Cypresshead Formation (Tc) consists of quartz sands ranging from very fine to very coarse, moderately to well sorted with common occurrences of quartz gravel. Clay is commonly present in very minor amounts and is generaly kaolinitic. Mica often occurs in minor percentages, particularly in finer grained sediments. Colors range from reddish orange in exposed sections to olive gray in the subsurface. The Cypresshead grades laterally down dip into the Nashua Formation which is a shelly clayey sand. The Cypresshead was deposited in a shallow nearshore marine setting and unconformably overlies the Hawthorn Group. It is unconformably overlain by undifferentiated sediments (QTu).

Miocene Hawthorn Group (Th) consists of marine sediments in central and eastern parts of the county. Two facies exist: Near the western limits of Hawthorn occurrence, the group is dominated by clays and clayey sands with minor carbonate and phosphate. To the east, sediments consist of more sand, less clay, and marine carbonate and phosphate. It is often exposed in sinkholes and scattered road cuts. Highly weathered Hawthorn Group sediments (Thw) occur on the Brooksville Ridge in the southwest corner of the county.

The Hawthorn Group unconformably overlies the Ocala Limestone and is overlain unconformably by the Cypresshead Formation (Tc) or undifferentiated sediments (QTu).

Tertiary Upper Eocene Ocala Group (To) consisting of very pure highly fossiliferous grainstones and packstones. Predominant fossils include large and small formainifera, mollusks, echinoids, and bryozoans occurring in the western part of the county. The Ocala limestone is highly karstified exhibiting dramatic local relief and is exposed in springs and quarries. The Ocala is overlain by unconformable Miocene Hawthorn Group sediments.

Baker County

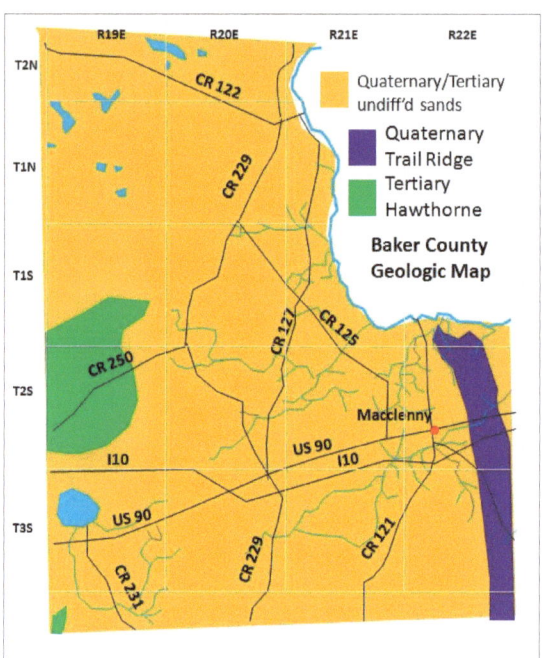

Figure 32. Baker County geologic map. Revised from Florida Geological Survey Open File Map No. 38.

Quaternary

Quaternary Trail Ridge (Qtr) quartz sands forming Trail Ridge and Baywood Promontory dunal sands, peat layers, heavy mineral placer deposits.

Quaternary/Tertiary

Undifferentiated Sands (QTu)- undifferentiated sands may contain some Cypresshead Formation with no general formations recognized. Lies on the Hawthorn Group or Ocala limestone. Often karstified with some of the karst features containing Hawthorn Group sediments.

Tertiary

Tertiary Hawthorn Group (Th)- sandy clays, silt, occasional carbonates locally containing phosphate.

Bradford/Union Counties

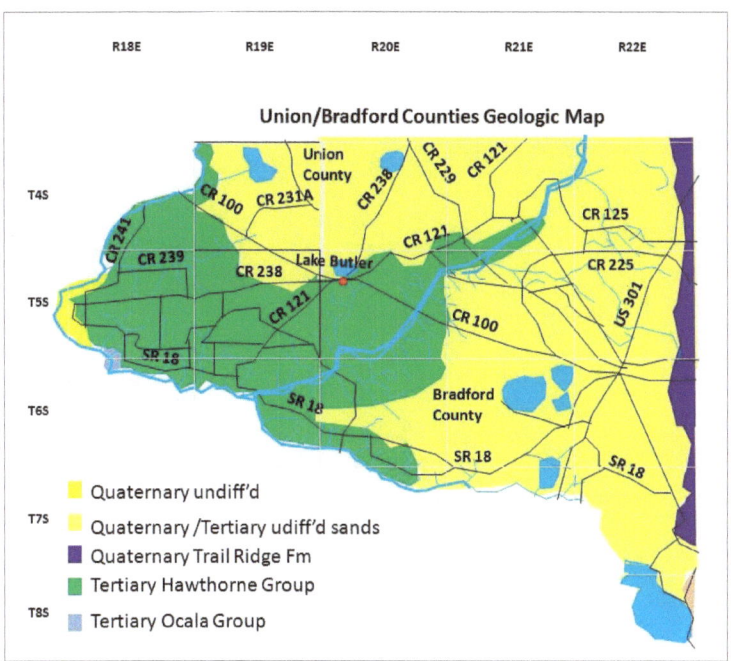

Figure 33. Bradford/Union Counties geologc map. Revised from Florida Geological Survey Open File Map No. 39.

Quaternary

Quaternary undifferentiated (Qu)- Undifferentiated quartz sands. May consist in part of reworked Citronelle Formation. Fine to coarse grained with varying percentages of silt and clay.

Quaternary Trail Ridge Formation (Qtr)- quartz sands forming Trail Ridge and Baywood Promontory dunal sands, peat layers, heavy mineral placer deposits.

Quaternary/Tertiary

Quaternary/Tertiary undifferentiated sands (QTu)- undifferentiated sands may contain some Cypresshead Formation with no general formations recognized. Lies on the Hawthorn Group or Ocala limestone. Often karstified with some of the karst features containing Hawthorn Group sediments.

Tertiary

Tertiary Hawthorn Group (Th)- sandy clays, silt, occasional carbonates locally containing phosphate.

Tertiary Ocala Group (To)- White to gray limestone, fossiliferous, moldic, varying from packstone to grainstone.

Brevard County

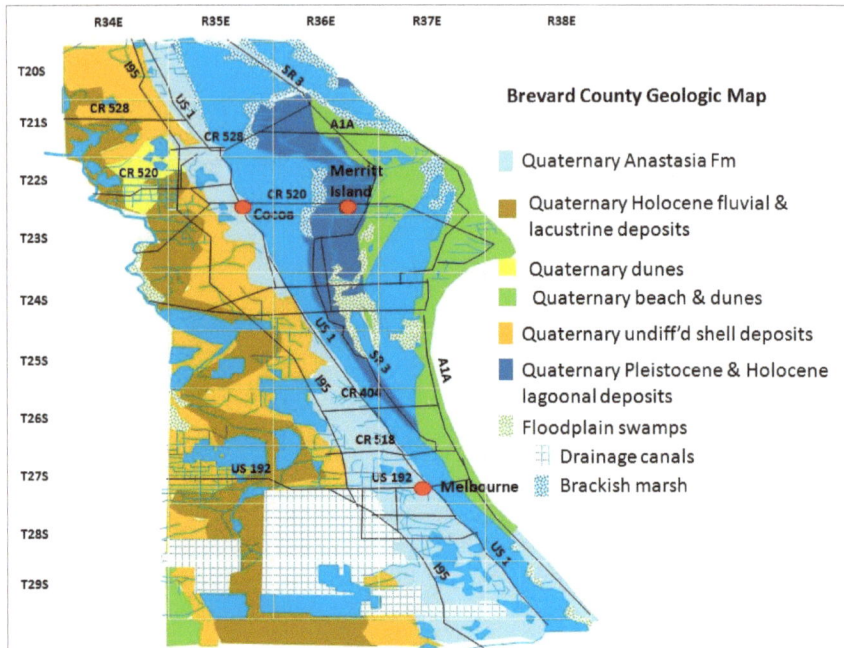

Figure 34. Brevard County geologic map. Revised from Florida Geological Survey Open File Map No. 49.

Quaternary

Quaternary Anastasia Formation (Qa)- Variably lithified coquina of shells and sands with unlithified fossiliferous sand.

Quaternary beach & dunes (Qbd)- Quartz sands with surface expression of beach ridges and dunes. A geomorphic unit on undifferentiated, often clean quartz sands. Generally fine to medium grained with no formations recognized. May contain shell.

Quaternary dunes (Qd)- Quartz sands with surface expression of dunes. A geomorphic unit on undifferentiated often clean quartz sands consisting of generally fine to medium grained.

Quaternary Pleistocene/Holocene (Qph/Qh)- Holocene sediments consisting of quartz sand with minor amounts of organic matter and clay associated with lagoonal deposits. Mostly beach and dune sands along present coastline. No formations recognized.

Quaternary Holocene fluvial and lacustrine (Qr)- sands, clays, marls, and peats in peninsular area only. No formations recognized.

Quaternary undifferentiated shell (Qsu)- shell beds, undifferentiated includes sediments previously placed in units primarily differentiated by the included fauna of the Caloosahatchee, Ft. Thompson, and Nashua Formations along with the Pinecrest Beds.

Citrus County

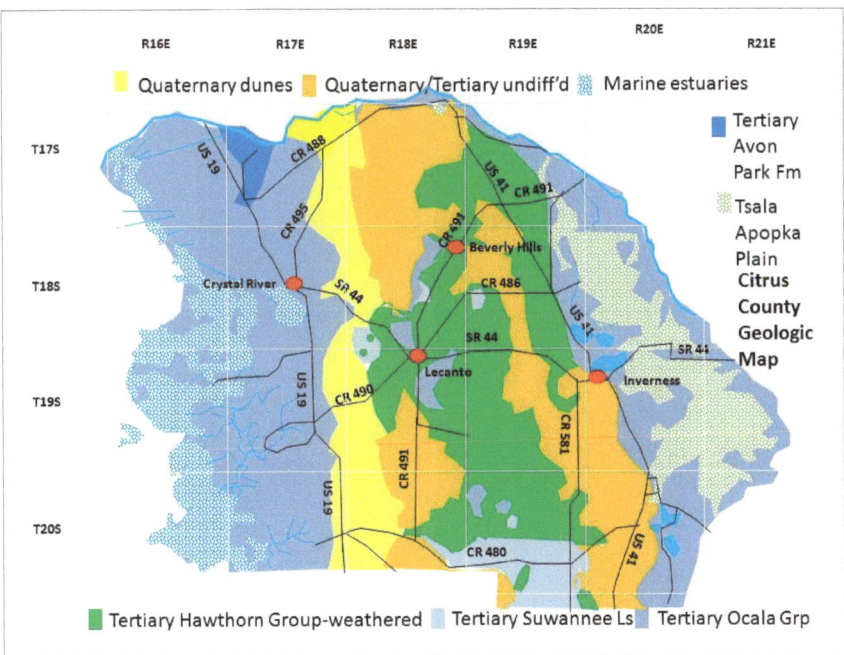

Figure 35. Citrus County geologic map. Revised from Florida Geological Survey Open File Map No. 10.

Quaternary

Quaternary dunes (Qd)- Quartz sands with surface expression of dunes. A geomorphic unit on undifferentiated often clean quartz sands consisting of generally fine to medium grained. Dune sands are present along the western flank of the Brooksville Ridge.

Quaternary/Tertiary

Quaternary/Tertiary undifferentiated sands (QTu)- undifferentiated sands may contain some Cypresshead Formation with no general formations recognized. Lies on the Hawthorn Group or Ocala limestone. Often karstified with some of the karst features containing Hawthorn Group sediments.

Tertiary

Tertiary Hawthorn Group weathered (Thw)- Weathered Hawthorn sediments form the bulk of the Brooksville Ridge in Citrus County. These sediments have been called the "Alachua Formation" in the past consisting primarily of red, orange, and light yellow variably clayey sands. These sediments are highly leached, clays are predominantly kaolin, and phosphate grains are generally removed.

Hard rock phosphate deposits are commonly associated with karst features along the eastern flank of the Brooksville Ridge. Weathered Hawthorn sediments are overlain by variable thicknesses of undiferentiated sediments.

Miocene Hawthorn Group (Th)- occurs as small outliers capping some of the highest hills in the central part of the county and in the southern part of the county. The Hawthorn consists of clays, clayey sands, and sandy, clayey limestone. All are phosphatic. The Hawthorn is overlain by weathered or reworked Hawthorn sediments or by undifferentiated sediments.

Oligocene Suwannee Limestone (Ts)- occurs within the Brooksville Ridge forming tops of some hills in the central portion of the county. The Suwannee consists of fossiliferous packstones and occasional grainstones slightly quartz rich sandy, variably recrystallized, commonly containing chert beds in the lower part of the unit. The Suwannee is overlain unconformably by the Hawthorn Group, by weathered and reworked Hawthorn sediments, or by undifferentiated sediments.

Upper Eocene Ocala Group limestone (To) is found at or near the land surface in the coastal lowlands in the western third of the county and in the Tsala Apopka Plain in the eastern portion of the county. The Ocala limestone consists of very pure, highly fossiliferous grainstones and packstones. Predominant fossils include large and small foraminifera, molluscs, echinoids, and bryozoans. The Ocala is unconformably overlain by the Suwannee Limestone, weathered Hawthorn sediments, unnamed Quaternary dune sands, or undifferentiated Quaternary sediments.

Tertiary Middle Eocene Avon Park Formation underlies the entire county but is exposed only in a small portion of northwestern Citrus County, consisting of dolostone and silt sized dolomite. The dolostone is often micro-moldic and highly recrystallized. The silt sized dolomite is often very soft to unconsolidated. The Avon Park is unconformably overlain by the Upper Eocene Ocala limestone.

Clay County

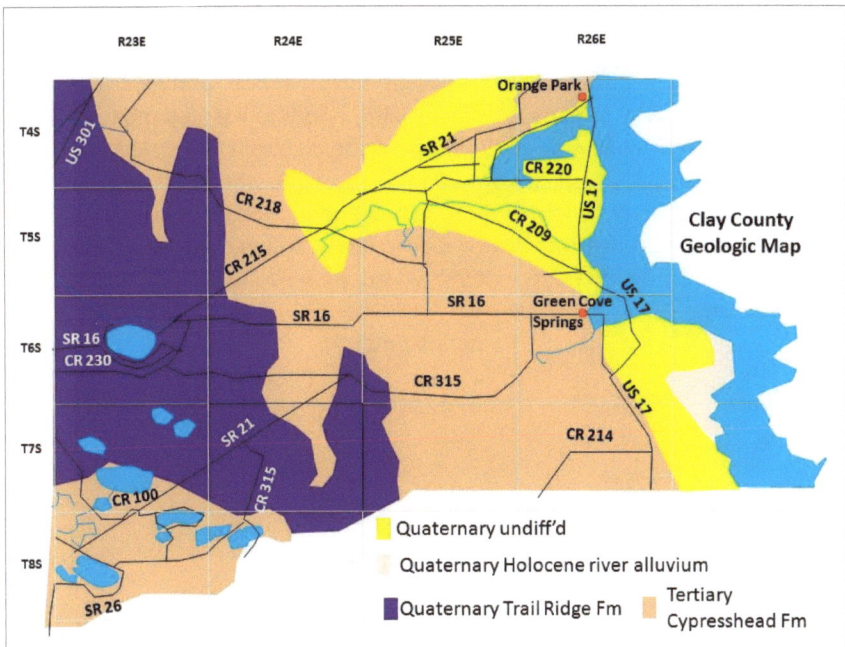

Figure 36. Clay County geologic map. Revised from Florida Geological Survey Open File Map No. 5

Quaternary

Quaternary undifferentiated (Qu)- occur in the eastern part of the county. Undifferentiated sediments are composed of sands, clayey sands, and clays occasionally containing limited numbers of mollusk shells. The sediment package may contain reworked Cypresshead Formation and lies unconformably on Nashua Formation of Hawthorn Group sediments. In localized areas, these sediments are very thin deposited though some fluvial reworking occurred.

Quaternary Holocene alluvium (Qr)- lowlands along the St. Johns River and some of its tributaries contain Holocene fluvial alluvial sediments composed of quartz sands, silt, clay, and marls poorly consolidated, variably sandy, clayey, shelly carbonate sediments. Peat and other organic rich sediments are often present. Unit is commonly present less than 10 feet in elevation.

Quaternary Trail Ridge Formation (Qtr)- occur in the western part of the county consisting of quartz sands fine to medium grained and moderately well sorted. Organic matter ranges from a finely divided organic matrix to peat beds and pieces of wood and common in the sands. The Trail Ridge sands contain economically important ore grade heavy mineral concentrations deposited as beach ridges and dunes.

Tertiary Late Pliocene Cypresshead Formation (Tc) consisting of quartz sands ranging from fine to very coarse moderately to well sorted with common occurrences of quartz gravel. Clay is commonly present in very minor amounts and is generally kaolinitic. Mica often occurs in minor percentages particularly in finer grained sediments. Colors range from reddish orange in exposed sections to olive gray in the subsurface. Faint, poorly preserved mollusk molds can be observed in exposures. The Cypresshead Formation grades laterally down dip into the Nashua Formation, a variably shelly, clayey sand. It unconformably overlies the Hawthorn Group and is overlain unconformably by the sands of Trail Ridge in western Clay County. Thin surficial sands and soil overlay it elsewhere. The Cypresshead was deposited in a shallow, nearshore, marine setting.

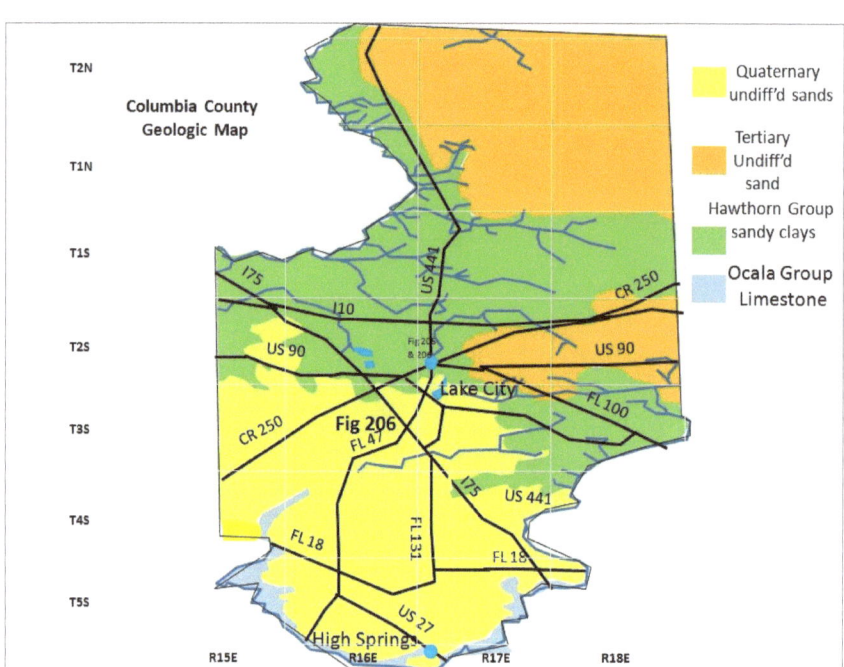

Figure 37. Columbia County geologic map. Revised from Florida Geological Survey Open File Map No. 37.

Quaternary

Quaternary undifferentiated sands (Qu)- undifferentiated surficial sands, clayey sands, clays, marls, and peats greater than 20 feet thick with no formations recognized.

Quaternary/Tertiary

Quaternary/Tertiary undifferentiated sand (QTu)- undifferentiated sands containing some Cypresshead Formation with no general formations recognized. Rests on Hawthorn Group or Ocala Group limestone. Occasionally karstified with some of the karst features containing Hawthorn Group sediments. Occurs at a higher elevation than Qu deposits.

Tertiary

Tertiary Hawthorn Group (Th)- consists of sandy clays, silt, and occasional carbonates with possible phosphate.

Tertiary Ocala Group (To)- white to gray fossiliferous moldic limestone variable between packstone and grainstone.

Duval County

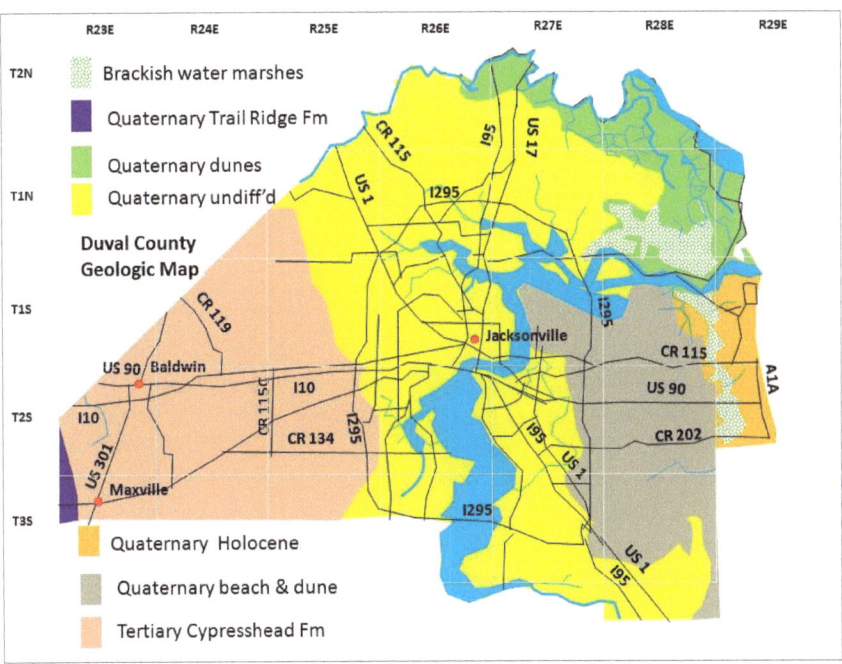

Figure 38. Duval County geologic map. Revised from Florida Geological Survey Open File Map No. 4.

Quaternary

Quaternary undifferentiated sand (Qu)- undifferentiated surficial sands, clayey sands, clays, marls, and peats greater than 20 feet thick with no formations recognized.

Quaternary dune deposits (Qd) - Quartz sands with surface expression of dunes. A geomorphic unit on undifferentiated often clean quartz sands consisting of generally fine to medium grained. Dune sands are present along the western flank of the Brooksville Ridge.

Quaternary Holocene (Qh)- Holocene sediments consisting of quartz sand with minor amounts of organic matter and clay associated with lagoonal deposits. Mostly beach and dune sands along present coastline. No formations recognized.

Quaternary beach & dune deposits (Qbd)- Quartz sands with surface expression of beach ridges and dunes. A geomorphic unit on undifferentiated, often clean quartz sands. Generally fine to medium grained with no formations recognized. May contain shell.

Quaternary Trail Ridge Formation (Qtr)- occurs in the exteme southwestern part of the county consisting of quartz sands fine to medium grained and moderately well sorted. Organic matter ranges from a finely divided organic matrix to peat beds and pieces of wood are common in the sands. The Trail Ridge sands contain economically important ore grade heavy mineral concentrations deposited as beach ridges and dunes.

Brackish Water Marshes- limited to the eastern part of the county where the St. Johns River enters the Atlantic Ocean.

Tertiary

Tertiary Cypresshead Formation (Tc)- consisting of quartz sands ranging from fine to very coarse moderately to well sorted with common occurrences of quartz gravel. Clay is commonly present in very minor amounts and is generally kaolinitic. Mica often occurs in minor percentages particularly in finer grained sediments. Colors range from reddish orange in exposed sections to olive gray in the subusrface. Faint, poorly preserved mollusk molds can be observed in exposures. The Cypresshead Formation grades laterally down dip into the Nashua Formation, a variably shelly, clayey sand. It unconformably overlies the Hawthorn Group and is overlain unconformably by the sands of Trail Ridge in southwestern Duval County. Thin surficial sands and soil overlay it elsewhere. The Cypresshead was deposited in a shallow, nearshore, marine setting.

Flagler County

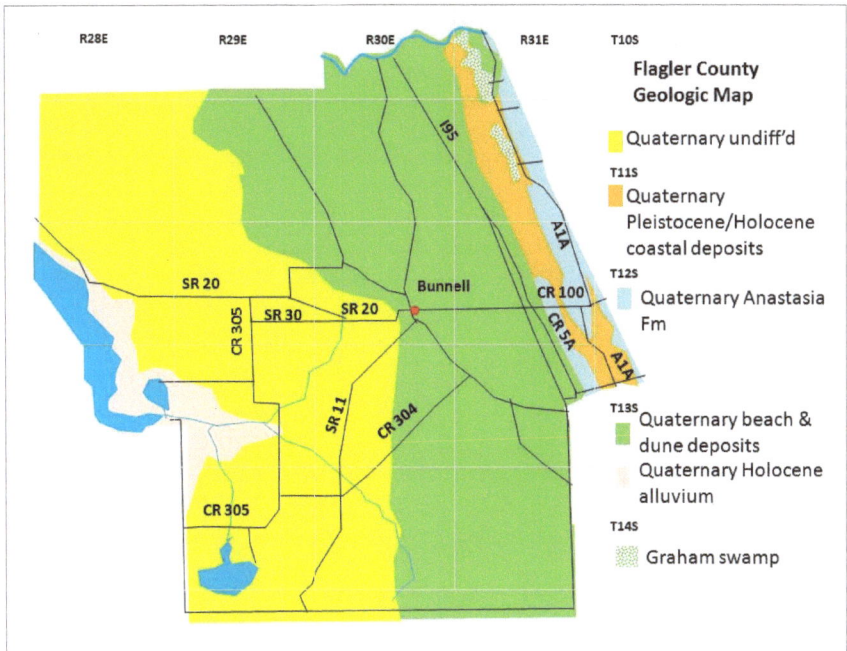

Figure 39. Flagler County geologic map. Revised from Florida Geological Survey Open File Map No. 7

Quaternary

Quaternary undifferentiated sands (Qu)- Flagler County is entirely covered by Quaternary deposits. Quaternary undifferentiated sediments occur in the western part composed of sands, clayey sands, and clays occasionally containing limited numbers of mollusk shells. Surficial conditions are often swampy and there are accumulations of organic matter. These sediments were deposited under marine conditions although some fluvial reworking occurred. This unit occurs above the Nashua Formation unconformably but underlies beach & dune (Qbd) deposits and Holocene alluvial deposits (Qr) where these deposits are recognized.

Quaternary beach & dune deposits (Qbd)- Beach ridges and dunes were deposited over much of the eastern half of Flagler County. Quartz sands comprising this unit are generally fine to medium grained moderately to well sorted and unfossiliferous. Sands contain significant percentages of heavy minerals. Deposits of organic material may form in swales between dunes and beach ridges. This unit may lie unconformably on undifferentiated Quaternary sediments or Nashua Formation. The surface expression of this unit controls the drainage pattern and streams follow the swales, coast parallel until encountering a larger stream perpendicularly crossing the ridges.

Quaternary Pleistocene/Holocene coastal deposits (Qph)- Along the present day coast, associated lagoons and coastal rivers and streams contain sediments referred to as undifferentiated Pleistocene and Holocene coastal deposits. These sediments are composed of sands, silts, and clays sometimes containing varying percentages of organic matter. Sands may contain heavy minerals and are poorly to well sorted depending on the depositional environment with varying amounts of fossils. Depositional environments include beach, marsh, and lagoonal sediments resting unconformably on older undifferentiated Quaternary sediments or Nashua Formation. Some occurrences of the Anastasia Formation may be included in this unit.

Quaternary Anastasia Formation (Qa)- The Anastasia Formation is the only lithostratigraphic unit recognized at or near the surface in the county occurring along the coast, on the barrier island, and Atlantic Coastal Ridge on the mainland. This unit consists of variably lithified coquina and mixtures of sand and shell. It is well exposed along the coast near Marineland and in pits between Flagler Beach and Bunnell.

Quaternary Holocene alluvium (Qr)- lowlands surrounding Crescent Lake and part of Haw Creek in westernmost Flagler County belong to the Holocene fluvial alluvial sediments composed of quartz sands, silt, clay, and marl poorly consolidated, variably sandy, clayey, shelly carbonate sediments. Peat and other organic rich sediments are often present. This unit is common below ten feet in elevation.

Gilchrist County

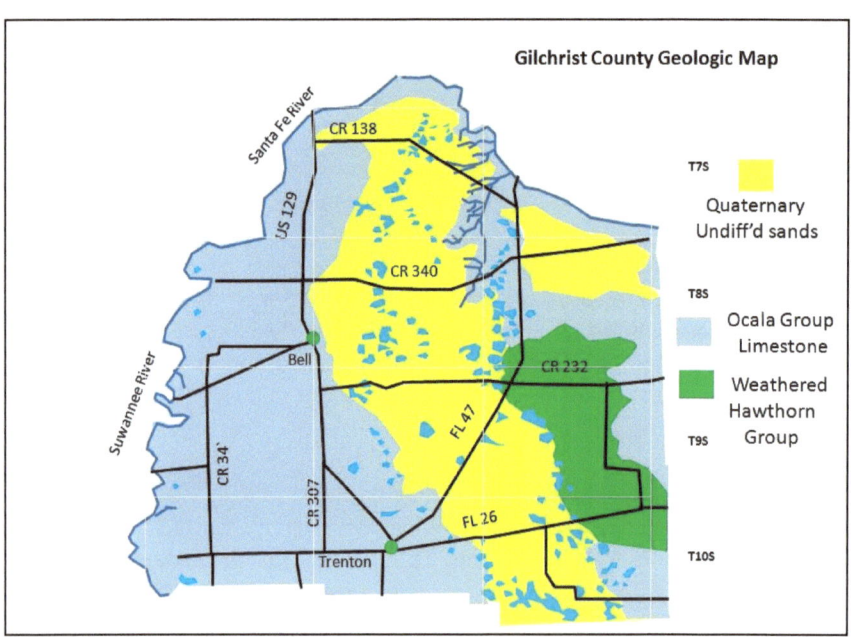

Figure 40. Gilchrist County geologic map. Revised from Florida Geological Survey Open File Map No. 36.

Quaternary

Quaternary undifferentiated sand (Qu)- undifferentiated surfical sands, clayey sands, clays, marls, and peats greater than 20 feet thick with no formations recognized.

Tertiary

Tertiary Hawthorn Group (Thw)- weathered Hawthorn Group sediments highly leached sands and clayey sands of the Brooksville Ridge.

Tertiary Ocala Group (To) limestone white to gray fossiliferous, moldic, ranging from packstone to grainstone.

Hernando County

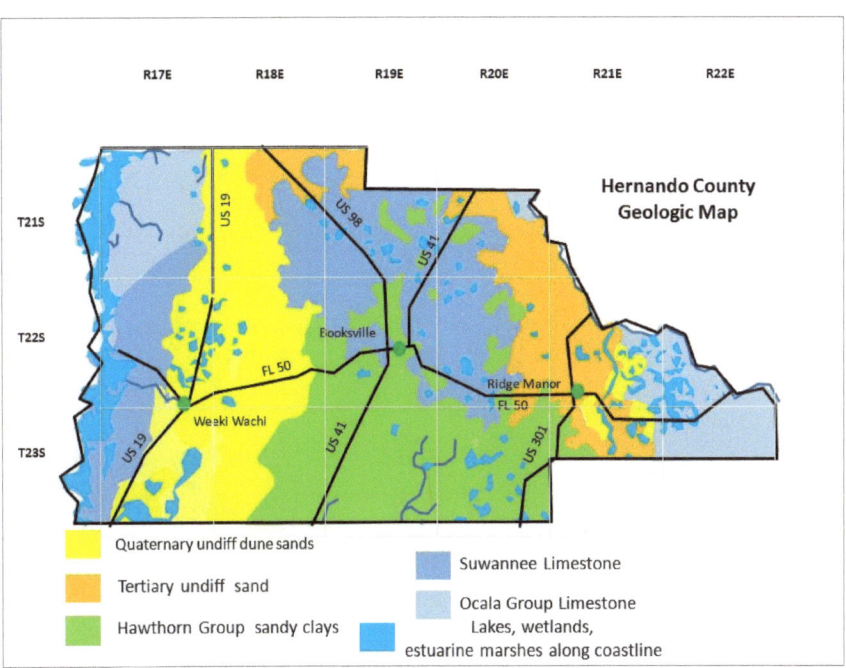

Figure 41. Hernando County geologic map. Revised from Florida Geological Survey Open File Map No. 41

Quaternary

Quaternary undifferentiated dune sands (Qd)- Quartz sands with a surface expression of dunes. A geomorphic unit on undifferentiated often clean quartz sands. Generally fine to medium grained.

Tertiary

Tertiary Hawthorn Group (Th)- Sandy clays, silt, occasional carbonates possibly containing phosphate.

Teritary Suwannee Limestone (Ts)- Marine limestone, fossiliferous, varies from poorly recrystallized to well recrystallized packstone to grainstone. Minor siliciclastic component.

Tertiary Ocala Group limestone, white to gray, fossiliferous, and moldic varies from packstone to grainstone.

Hillsborough County

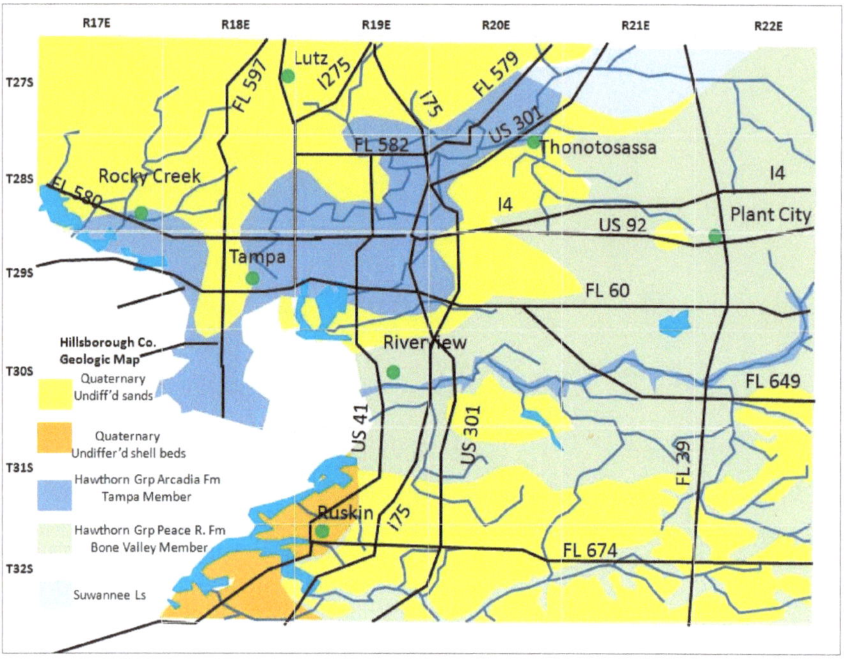

Figure 42. Hillsborough County geologic map. Revised from Florida Geological Survey Open File Map No. 45

Quaternary

Quaternary undifferentiated sands (Qu)- undifferentiated surficial sands, clayey sands, clays, marls, and peats greater than 20 feet thick with no formations recognized.

Quaternary undifferentiated shell beds (Qsu)- shell beds, undifferentiated including sediments previously placed in the Caloosahatchie, Ft. Thompson, and Nashua Formations including the Pinecrest Beds.

Hawthorn Group Arcadia Formation Tampa Member (That)- consists predominantly of limestone with subordinate dolostone, quartz sands, and clays. Limestone contains variable quartz sand and clay but contains little or no phosphate.

Tertiary Hawthorn Group Peace River Formation Bone Valley Member consisting of pebble or gravel sized phosphate fragments and sand sized phosphate grains in a matrix of quartz sand and clay. Percentages of variable components are highly variable.

Tertiary Suwannee Limestone (Ts)- Marine limestone, fossiliferous varies from poorly recrystallized to well recrystallized packstone to grainstone with minor siliciclastic component.

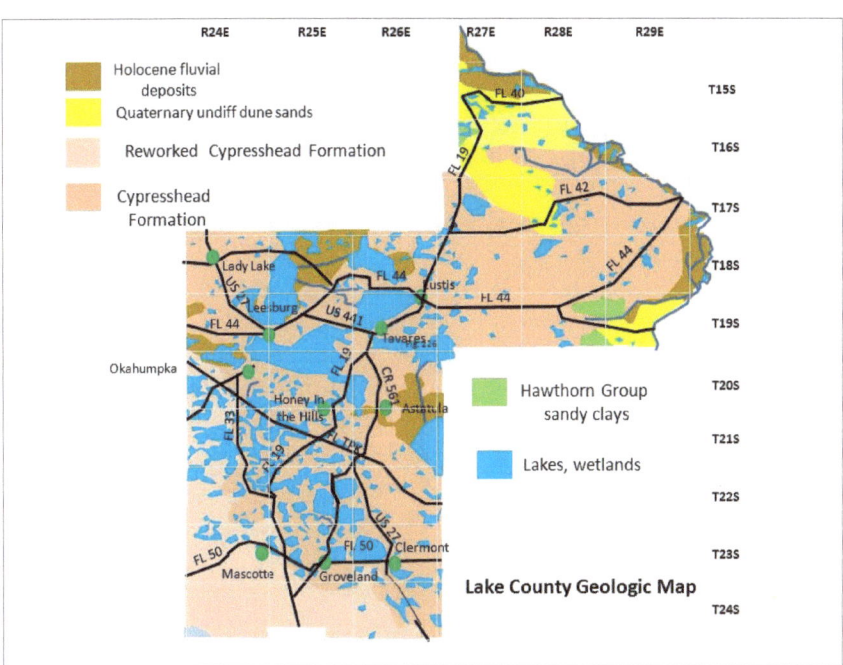

Figure 43. Lake County geologic map. Revised from Florida Geological Survey Open File Map No. 9

Quaternary

Quaternary dune sands (Qd)- Composed of sands, clayey sands, and clays possibly containing limited numbers of mollusk shells. Surficial conditions are often swampy and there are accumulations of organic matter. Sediments were deposited under marine conditions although fluvial reworking occurred. Reworked Cypresshead Formation are thought to be present. Unit rests unconformably on the Cypresshead Formation or undifferentiated Pleistocene sediments.

Quaternary dune sands (Qd)- In northeastern Lake County this unit extends into Marion County. Karst modification of the dunes is often present.

Quaternary alluvium (Qr) – lowlands surrounding the St. Johns River and tributaries including lakes in the Central Valley are Holocene fluvial and lacustrine sediments composed of quartz sands, silt, clay, and marl poorly consolidated, variably sandy, clayey, shelly, carbonate sediments. Peat and other organic rich sediments are often present.

Tertiary

Tertiary Late Pliocene Cypresshead Formation (Tc) consists of quartz sands ranging from fine to very coarse moderately to well sorted with common occurences of quartz gravel. Clay is commonly present in very minor amounts and is particularly in the finer grained sediments. Colors range from reddish orange in exposed sediments to olive gray in subsurface sediments. Cypresshead sediments grade laterally down dip into the Nashua formation. The Cypresshead was deposited in a shallow nearshore marine setting and unconformably overlies the Hawthorn Group.

Tertiary Late Miocene Hawthorn Group (Th)- Hawthorn sediments are found near the surface in low lying areas associated with spring vents and runs along the bottoms of some rivers and streams in northeastern Lake County. These sediments can be seen in the area of Seminole and Rock Springs. Sediments consist of admixtures of sand, silt, clay, and carbonate with varying percentages of phosphate present. These sediments were deposited under marine conditions. The Hawthorn Group unconformably overlies the Eocene Ocala Group and is overlain by the Cypresshead Formation.

Levy County

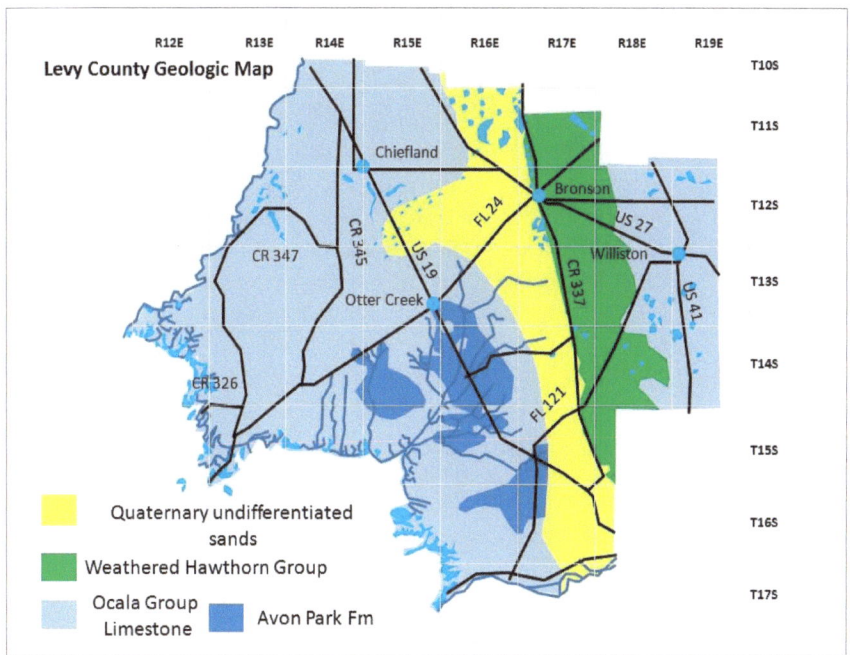

Figure 44. Levy County geologic map. Revised from Florida Geological Survey Open File Map No. 11

Quaternary

Quaternary undifferentiated sands (Qu)- Undifferentiated Quaternary sediments are present at the surface throughout most of the county. They are thin to absent throughout most of the area where Ocala Group or Avon Park formations are present. Undifferentiated sediments achieve thicknesses greater than 20 feet only in karstic features and in the Waccasassa Flats, a north to south band on the western flank of the Brooksville Ridge. Undifferentiated sediments consist of terrace sands, poorly developed dune fields, unnamed Pleistocene limestones and marls and Holocene peats and fluvial deposits.

Tertiary

Tertiary Hawthorn Group weathered deposits (Thw)- sediments comprising the Brooksville Ridge which trends north to south through eastern Levy County. These sediments were referred to as the Alachua Formation in the past consisting of primarily red and orange variably clayey sands. Clays are predominantly kaolin and the sediments were highly leached with phosphate grains being removed. Hard rock phosphate deposits are associated with karst features along the flanks of the ridge. Weathered Hawthorn Group sediments are overlain by variable thicknesses of undifferentiated sediments (Qu).

Tertiary Miocene Hawthorn Group (Th)- is present in Levy County only as a small outlier along the eastern edge of the county in the vicinity of Williston. The Hawthorn in this area consists of clayey sands and sandy clays with variable percentages of phosphate.

Tertiary Upper Eocene Ocala Group limestone unconformably overlies the Avon Park formation and is found at or near the land surface in the western half and along the eastern edge of the county. The Ocala limestone consists of very pure highly fossiliferous grainstones and packstones containing abundant foraminifera, mollusks, and echinoids in the vicinity of the Avon Park Formation exposures where the Ocala limestone is very thin. Weathered reworked Hawthorn Group sediments overlie the Ocala Group along with undifferentiated Quaternary sediments (Qu).

Tertiary Middle Eocene Avon Park Formation (Tap)- The Avon Park formation consists of variably textured fossiliferous limestone and dolostone containing small quantities of organic material and peat. The Avon Park is exposed in the central and south central portions of the county overlain by the Upper Eocene Ocala Group limestone.

Marion County

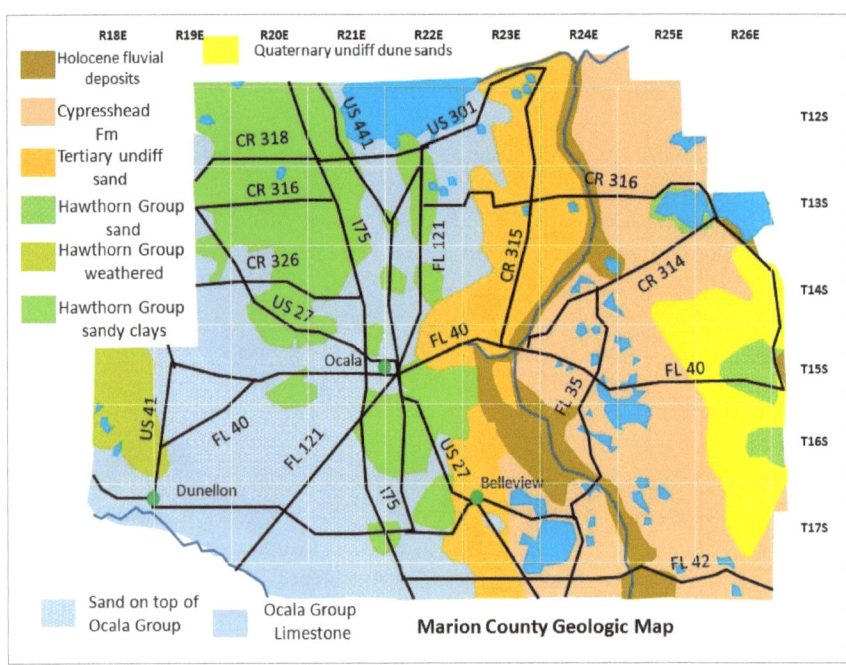

Figure 45. Marion County geologic map. Revised from Florida Geological Survey Open File Map No. 13.

Quaternary

Quaternary Holocene deposits (Qr)- Holocene fluvial sands, clays, marls, and peats with no formations recognized.

Quaternary dune sands (Qd)- Dune sands often extensively modified by karstic processes with formations recognized. Lies on top of Hawthorn Group or Cypresshead Formation sediments.

Quaternary/Tertiary

Quaternary/Tertiary undifferentiated sands (QTu)- undifferentiated sands may contain some Cypresshead Formation with generally no distinct formations recognized. Lies on Hawthorn Group or Ocala limestone. Often karstified with some of the karst features containing Hawthorn Group sediments.

Tertiary

Tertiary Cypresshead Formation (Tc)- Quartz sands with minor clay occurring on top of Hawthorn Group.

Tertiary Hawthorn Group (Th)- Highly phosphatic clays, quartz sands, and carbonates.

Tertiary Hawthorn Group weathered sediments (Thw)- Highly leached sediments on the Brooksville Ridge in the southwestern part of Levy County.

Tertiary sands on top of Ocala Group (Ths)- Sediments covered by a thin sequence of sand due to high water tables on top of Ocala limestone.

Tertiary Ocala Limestone (To)- Very pure limestone with occasional chert, often highly karstified with karst features infilled by Hawthorn Group sediments and or quartz sands.

Nassau County

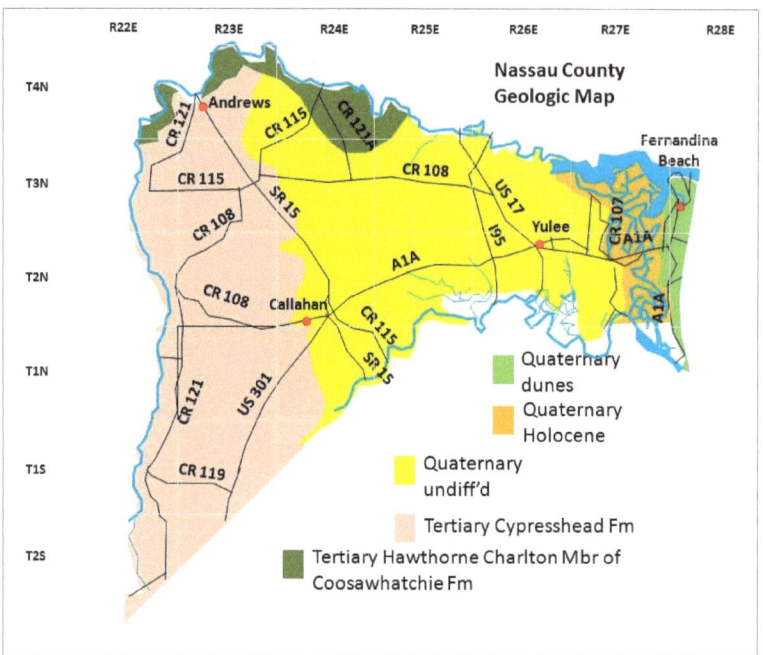

Figure 46. Nassau County geologic map. Revised from Florida Geological Survey Open File Map No. 3

Quaternary

Quaternary dune sands (Qd)- Sand dunes form Amelia Island on the eastern coast of Nassau County consisting of well sorted quartz sands comprising the dunes which may contain shell fragments. The sands often contain 5 to 10 percent heavy minerals. The unit rests on top of undifferentiated Pleistocene and Holocene coastal deposits.

Quaternary Holocene deposits (Qph/Qh)- Along the present day coast, the associated lagoons, coastal rivers, and streams are sediments referred to as undifferentiated Pleistocene and Holocene coastal deposits. The sediments are composed of sands, silts, and clays sometimes containing varying percentages of organic matter. The sands may contain mica and heavy minerals. Sands are poorly to well sorted depending on the depositional environment. These include beach, marsh, and lagoonal sediments resting unconformably on older undifferentiated Quaternary sediments or Nashua Formation.

Quaternary undifferentiated sediments (Qu)- occur in the central and eastern portions of the county composed of sands, clayey sands, and clays occasionally containing limited numbers of fossils. This sediment package may contain reworked Cypresshead Formation and rests unconformably on Nashua Formation or Hawthorn Group sediments.

Locally, these sediments are very thin and were deposited under marine conditions although some fluvial reworking occurred.

Tertiary

Tertiary Late Pliocene Cypresshead Formation (Tc)- consists of quartz sands ranging from fine to very coarse moderately to well sorted with common occurrences of quartz gravel. Clay is commonly present in very minor amounts and is generally kaolinitic. Mica often occurs in minor percentages particularly in finer grained sediments. Colors range from reddish orange in exposed section to olive gray in the subsurface. Faint poorly preserved mollusk molds are visible in exposures. The Cypresshead Formation grades laterally down dip into the Nashua Formation which is a shelly, clayey sand. It unconformably overlies the Hawthorn Group and is overlain unconformably by the Trail Ridge sands in westernmost Duval County and thin surficial sands elsewhere. The Cypresshead was deposited in a shallow nearshore marine setting.

Tertiary Charlton Member of the Coosawhatchie Formation which belongs to the Hawthorn Group (Thcc)- Sediments are exposed along the St. Marys River consisting of interbedded carbonate limestone and dolostone, and clay beds. The carbonates are slightly sandy and clayey often containing abundant molds and casts of mollusks. Clays are silty and calcareous to dolomitic.

Orange County

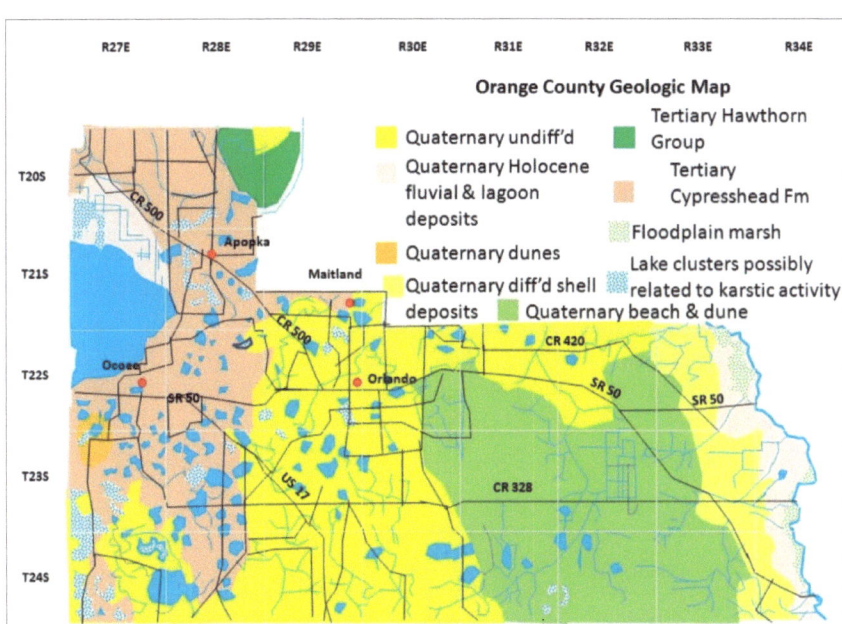

Figure 47. Orange County geologic map. Revised from Florida Geological Survey Open File Map No. 47.

Quaternary

Quaternary beach & dune deposits (Qbd)- Quartz sands with surface expressions of beach ridges and dunes. A geomorphic unit on undifferentiated often clean quartz sands. Generally fine to medium grained with no formations recognized. May contain shell.

Quaternary dune sands (Qd)- Quartz sands with surface expression of beach ridges and dunes. A geomorphic unit on undifferentiated often clean quartz sands. Generally fine to medium grained with no formations recognized.

Quaternary Holocene fluvial and lagoonal deposits (Qr)- Holocene fluvial and lacustrine sands, clays, marls, and peats. No formations recognized occurs in peninsular areas only.

Quaternary undifferentiated shell (Qsu)- Shell beds, undifferentiated includes sediments previously placed in the Caloosahatchie, Ft. Thompson, and Nashua Formation including the Pinecrest Beds.

Quaternary undifferentiated sands (Qu)- Undifferentiated surficial sands, clayey sands, clays, marls, and peats greater than 20 feet thick with no formations recognized.

Floodplain marsh – occurs in the easternmost part of the county adjacent to the St. Johns River.

Karstic lake clusters – occurs in the western third of the county where small clusters of lakes appear to be related to sinkhole activity in the underlying limestone.

Tertiary

Tertiary Cypresshead Formation (Tc)- Cypresshead Formation quartz sands with minor clay. Rests on the Hawthorn Group.

Tertiary Hawthorn Group (Ths)- Hawthorn Group sediments near surface exposed in stream runs and springs under generally swampy conditions. Surface sediments may be greater than 20 feet thick away from streams and springs.

Osceola County

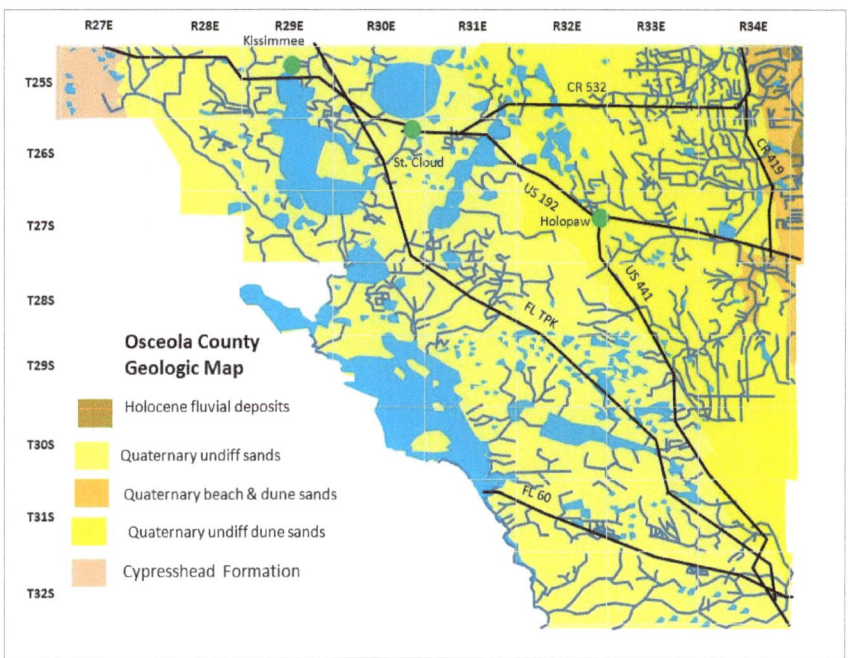

Figure 48. Osceola County geologic map. Revised from Florida Geological Survey Open File Map No. 48

Quaternary

Quaternary beach and dune sands (Qbd)- Quartz sands with surface expressions of beach ridges and dunes. A geomorphic unit on undifferentiated often clean quartz sands. Generally fine to medium grained with no formations recognized. May contain shell.

Quaternary Holocene deposits (Qr)- Holocene fluvial and lacustrine sands, clays, marls, and peats. No formations recognized and applies to peninsular areas only.

Quaternary undifferentiated shell beds (Qsu- pale yellow on map)- Shell beds, undifferentiated includes sediments placed in the Caloosahatchie, Ft. Thompson, and Nashua Formation including Pinecrest Beds.

Quaternary undifferentiated sands (Qu)- Undifferentiated surficial sands, clayey sands, clays, marls, and peats greater than 20 feet thick with no formations recognized.

Tertiary

Tertiary Cypresshead Formation (Tc)- Cypresshead Formation quartz sands with minor clay. Rests on the Hawthorn Group.

Pasco County

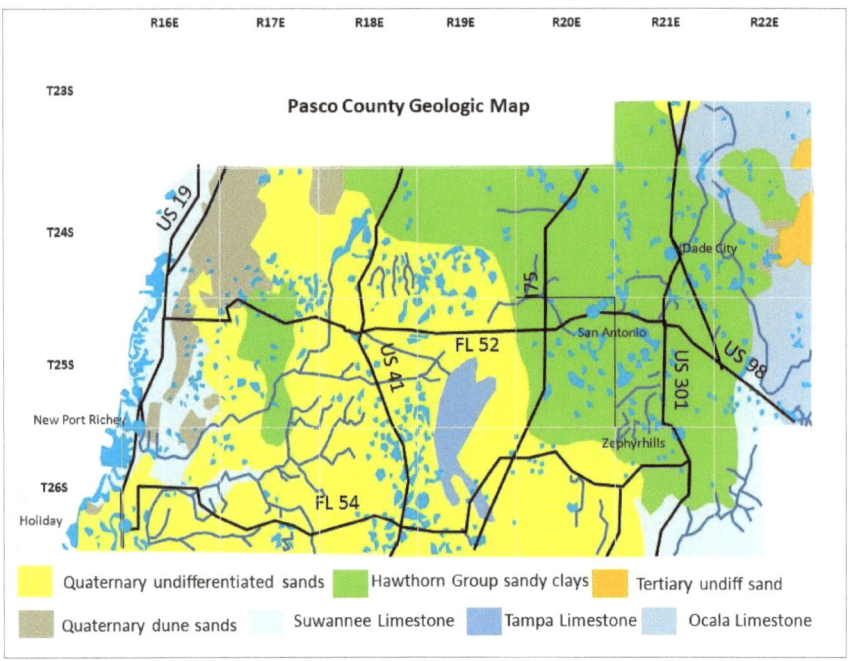

Figure 49. Pasco County geologic map. Revised from Florida Geological Survey Open File Map No. 42

Quaternary

Quaternary dune sands (Qd)- Quartz sands with surface expression of beach ridges and dunes. A geomorphic unit on undifferentiated often clean quartz sands. Generally fine to medium grained with no formations recognized.

Quaternary undifferentiated sands (Qu)- Undifferentiated surficial sands, clayey sands, clays, marls, and peats greater than 20 feet thick with no formations recognized.

Quaternary/Tertiary

Quaternary/Tertiary undifferentiated sands (QTu- listed as Tertiary undifferentiated sand on map)- Undifferentiated sands containing some Cypresshead Formation. Generally no formations are recognized resting on Hawthorn Group or Ocala Group limestone. Often karstified with some karst features containing Hawthorn Group sediments.

Tertiary

Tertiary Hawthorn Group (Th)- Sandy clays, silt, occasional carbonates may contain phosphate and is variably weathered.

Hawthorn Group Arcadia Formation Tampa Member (That)- Generally sandy limesotne with very low to no phosphate content. Variably fossiliferous.

Tertiary Suwannee Limestone (Ts)- Suwannee Limestone fossiliferous limestone varies from poorly recrystallized to well recrystalled packstone to grainstone with minor siliciclastic component.

Tertiary Ocala Group limestone (To)- White to gray, fossiliferous moldic limestone. Varies from wackstone to grainstone.

Pinellas County

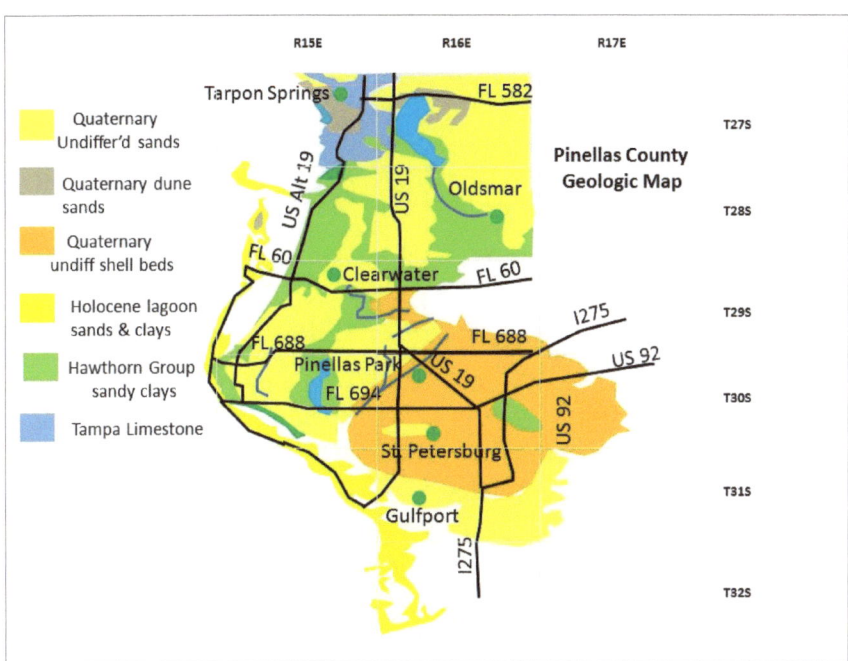

Figure 50. Pinellas County geologic map. Revised from Florida Geological Survey Open File Map No. 44

Quaternary

Quaternary dune sands (Qd)- Quartz sands with surface expression of beach ridges and dunes. A geomorphic unit on undifferentiated often clean quartz sands. Generally fine to medium grained with no formations recognized.

Quaternary Holocene sediments (Qh)- Holocene sediments consisting of quartz sand with minor amounts of organic matter and clay associated with lagoonal deposits. Mostly beach and dune sands along the present day coastline with no formations recognized.

Quaternary undifferentiated shell beds (Qsu)- Shell beds, undifferentiated includes sediments placed in the Caloosahatchie, Ft. Thompson, and Nashua Formation including Pinecrest Beds.

Quaternary undifferentiated sands (Qu)- Undfferentiated surficial sands, clayey sands, clays, marls, and peats greater than 20 feet thick with no formations recognized.

Tertiary

Tertiary Hawthorn Group (Th)- Clays, sandy clays, silt occasional carbonates which are usually phosphatic.

Tertiary Hawthorn Group Arcadia Formation Tampa Member (That)- Consists predominantly of limestone with subordinate dolostone, quartz sands, and clays. Limestone contains various quartz sand and clay but contains little or no phosphate.

Polk County

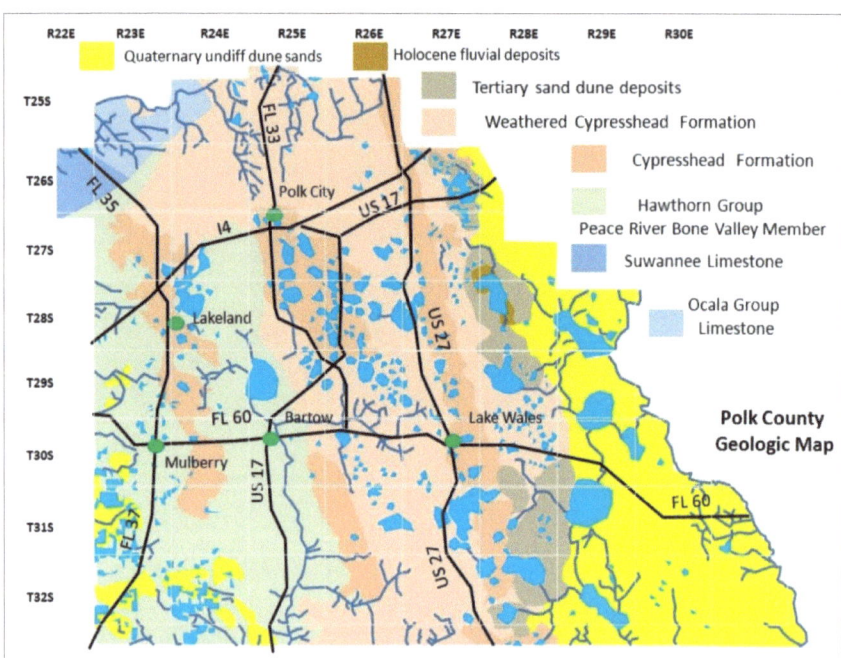

Figure 51. Polk County geologic map. Revised from Florida Geological Survey Open File Map No. 46

Quaternary

Quaternary Holocene fluvial and lacustrine (Qr)- Holocene fluvial and lacustrine sands, clays, marls, and peats. No formations recognized and applies to peninsular areas only.

Quaternary undifferentiated (Qu)- Undfferentiated surficial sands, clayey sands, clays, marls, and peats greater than 20 feet thick with no formations recognized.

Quaternary weathered Cypresshead Formation (Quc)- undifferentiated sands reqworked from Cypresshead Formation with isolated occurrences of Cypresshead formation present.

Quaternary/Tertiary

Quaternary/Tertiary (QTu- shown as Tertiary undifferentiated on map)- Quartz sands with surface expression of dunes. A geomorphic unit on undifferentated often clean quartz sands. Generally fine to medium grained.

Tertiary

Tertiary Cypresshead Formation (Tc)- Quartz sands with minor clay resting on Hawthorn Group.

Tertiary Hawthorn Group Peace River Formation Bone Valley Member (Thpb)- Pebble or gravel sized phosphate fragments and sand sized phosphate grains in a matrix of quartz sand and clay. Percentages of the varous components are variable.

Tertiary Suwannee Limestone (Ts)- Marine limestone, fossiliferous, moldic varying from poorly to well recrystallized packstone to grainstone with minor siliciclastics.

Tertiarty Ocala Group (To)- White to gray, fossiliferous, moldic limestone. Varies from wackstone to grainstone.

Putnam County

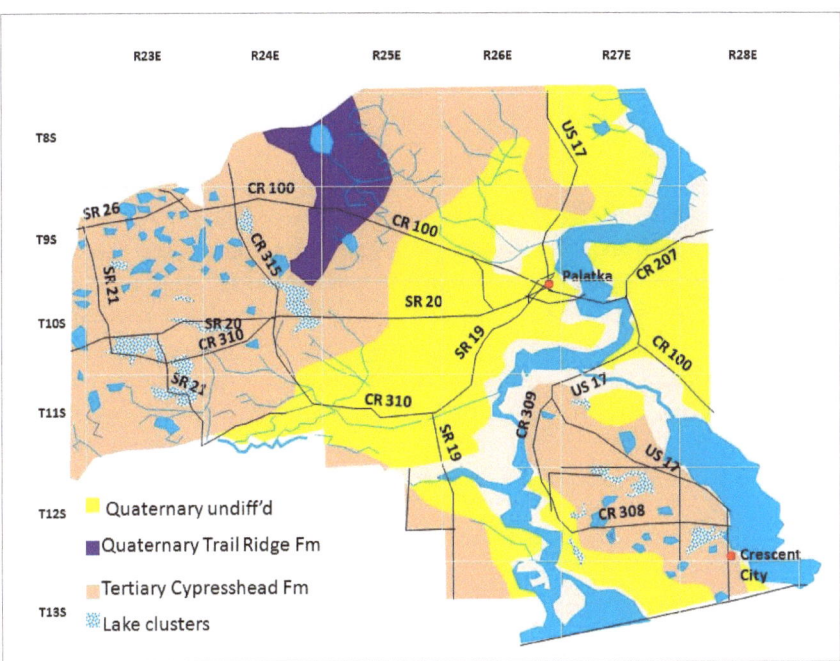

Figure 52. Putnam County geologic map. Revised from Florida Geological Survey Open File Map No. 6

Quaternary

Quaternary Holocene sediments (Qr- shown as tan colors along the St. Johns River on map)- Lowlands along the St. Johns River and some of its tributaries including Crescent Lake and the Oklawaha River belong to Holocene fluvial sediment alluvium composed of quartz sands, silt, clay, and marls that are poorly consolidated, variably sandy, clayey, and shelly carbonate sediments. Peat and the rorganic rich sediments are often present. Occurs below ten feet in elevation.

Quaternary undifferentiated sediments (Qu)- occur in widespread undifferentiated sediments composed of sands, clayey sands, and clays occasionally containing limited numbers of mollusk shells and organic deposits. This sediment package may contain reworked Cypresshead Formation and lies unconformably on Nashua Formation or Hawthorn Group sediments. These sediments were deposited under marine conditions though fluvial reworking occurred.

Quaternary Trail ridge formation (Qtr)- Overlying the Cypresshead Formation in the north central portion of the county are Quaternary Pleistocene sands forming trail Ridge and associated features. Quartz sands are generally fine to medium grained and moderately to well sorted. Organic matter ranges from a finely divdided organic matrix to peat beds and pieces of wood are common in the Trail Ridge sands. Trail Ridege sands contain economically important ore grade heavy mineral concentrations. Sands were deposited as beach ridges and dunes.

Tertiary Late Pliocene Cypresshead Formation (Tc)- consisting of quartz sands ranging from fine to very coarse, moderately to well sorted with common occurences of quartz gravel. Clay is commonly present in very minor amounts and is generally kaolinitic. Mica often occurs in minor percentages particularly in finer grained sediments. Colors range from reddish orange in exposed section to olive gray in the subsurface. Faint, poorly preserved mollusk molds are present in exposures. The Cypresshead grades laterally down dip into the Nashua Formation, a variable shelly clayey sand. It unconformably overlies the Hawthorn Group and is overlain by sand of the Trail Ridge unconformably in western Clay County and thin surficial sands and soils elsewhere. The Cypresshead was deposited in a shallow nearshore marine setting.

St. Johns County

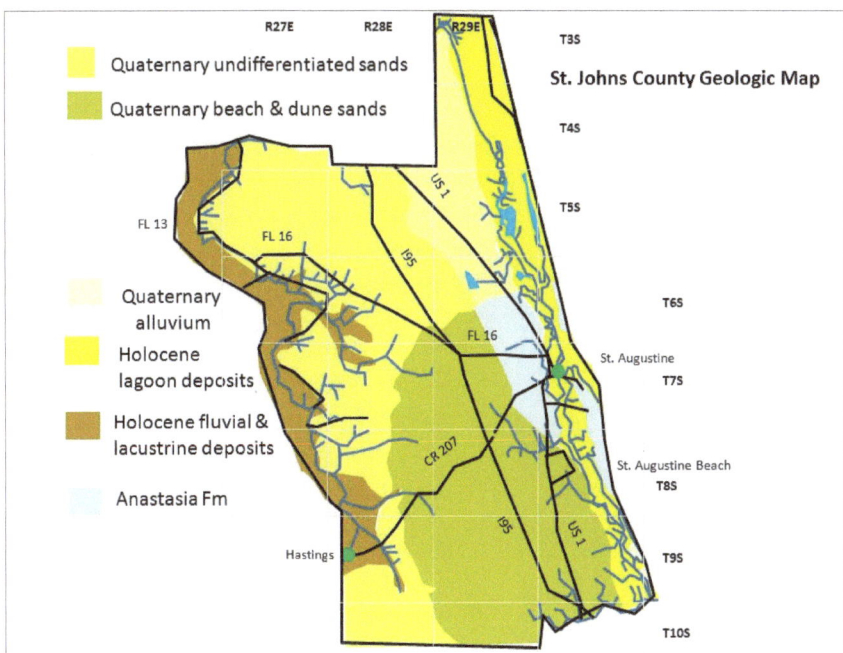

Figure 53. St. Johns County geologic map. Revised from Florida Geological Survey Open File Map No. 68

Quaternary

Quaternary Anastasia Formation (Qa)- Variably lithified coquina of shells and sands with unlithified fossiliferous sand.

Quaternary beach & dune deposits (Qbd)- Quartz sands with surface expression of beach ridges and dunes. A geomorphic unit on undfferentiated clean quartz sands. Generally fine to medium grained with no formations recognized. May contain shell.

Quaternary Holocene sediments (Qph)- Holocene sediments consisting of quartz sand with minor amounts of organic matter and clay associated with lagoonal deposits. Mostly beach and dune sands along present coastline with no formations recognized.

Quaternary Holocene fluvial and lacustrine (Qr)- Holocene fluvial and lacustrine sands, clays, marls, and peats. No formations recognized and is applied to peninsular areas only.

Quaternary undifferentiated sands (Qu)- Undifferentiated surficial sands, clayey sands, clays, marls, and peats greater than 20 feet thick with no formations recognized.

Seminole County

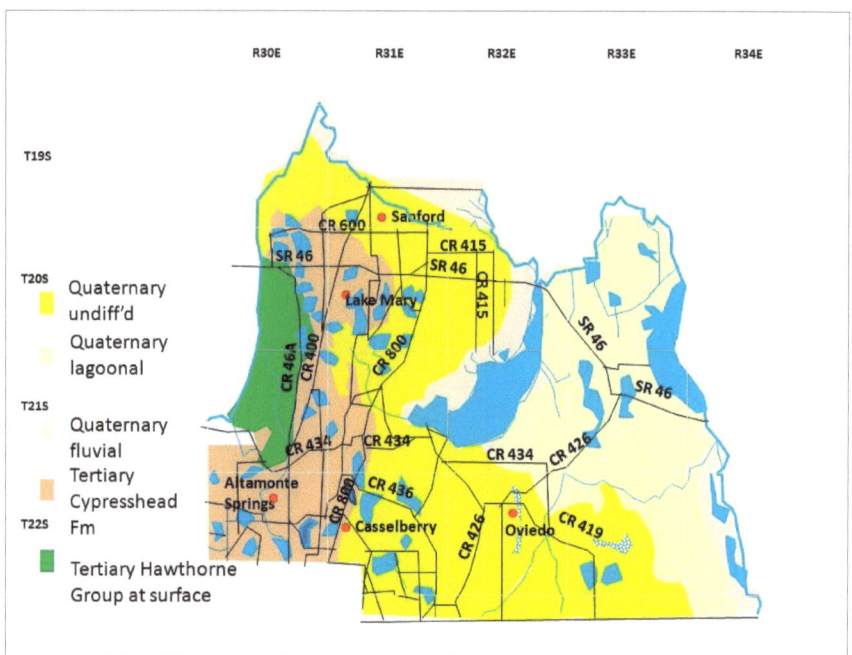

Figure 54. Seminole County geologic map. Revised from Florida Geological Survey Open File Map No. 43

Quaternary

Quaternary Holocene fluvial (Qr)- Holocene fluvial sands, clays, marls, and peats with no formation recognized but may be similar to Quaternary alluvial designates.

Quaternary unidfferentiated (Qu)- undifferentiated surficial sands, clayey sands, clays, marls, and peats greater than 20 feet thick with no formations recognized.

Quaternary lagoonal deposits (Qul)- undifferentiated sands with variable amounts of clays, occasionally shelly. Sand appear to be lagoonal deposits with peats formed occasionally at the surface.

Tertiary

Tertiary Cypresshead Formation (Tc)- Quartz sands with minor clay resting on top of Hawthorn Group.

Tertiary Hawthorn Group near surface (Ths)- Exposed in stream runs and springs. Generaly swampy conditions. Surface sediments may exceed 20 feet thick away from streams and springs.

Sumter County

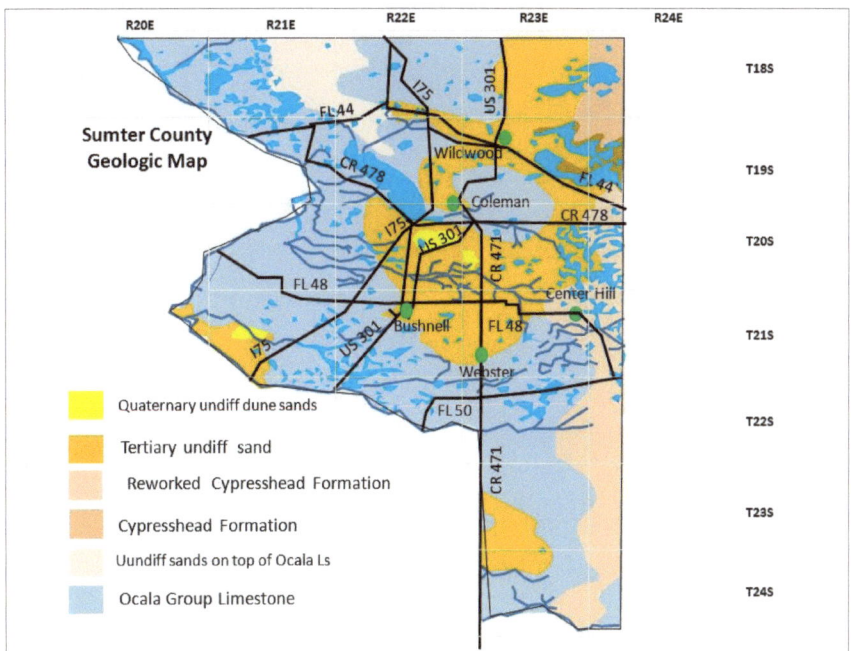

Figure 55. Sumter County geologic map. Revised from Florida Geological Survey Open File Map No. 40

Quaternary

Quaternary dunes (Qd)- Quartz sands with surface expression of dunes. A geomorphic unit on undifferentiated often clean quartz sands generally fine to medium grained.

Quaternary Holocene fluvial and lacustrine (Qr)- Holocene fluvial sands, clays, marls, and peats with no formation recognized but may be similar to Quaternary alluvial designates.

Quaternary Cypresshead reworked sediments (Quc)- Undifferentianted sands reworked from Cypresshead Formation with isolated occurrences of Cypresshead Formation sediments.

Quaternary/Tertiary

Quaternary/Tertiary undifferentiated sands (QTu- labeled as Tertiary undifferentiated sands on map)- Undifferentaited sands may contain some Cypresshead Formation with no formations recognized. Rests on top of Hawthorn Group or Ocala Group limestone. Often karstified with some of the karst features containing Hawthorn Group sediments.

Quaternary/Tertiary undifferentated sands (QTuk)- undifferentiated sands lying on highly karstic Ocala Group limestgone. Karstic features may contain Hawthorn Group sediments.

Tertiary

Tertiary Cypresshead Formation (Tc)- Quartz sands with minor clay resting on top of Hawthorn Group sediments.

Tertiary Ocala Group limestone (To)- White to gray fossiliferous marine limestone variable from packstone to grainstone.

Volusia County

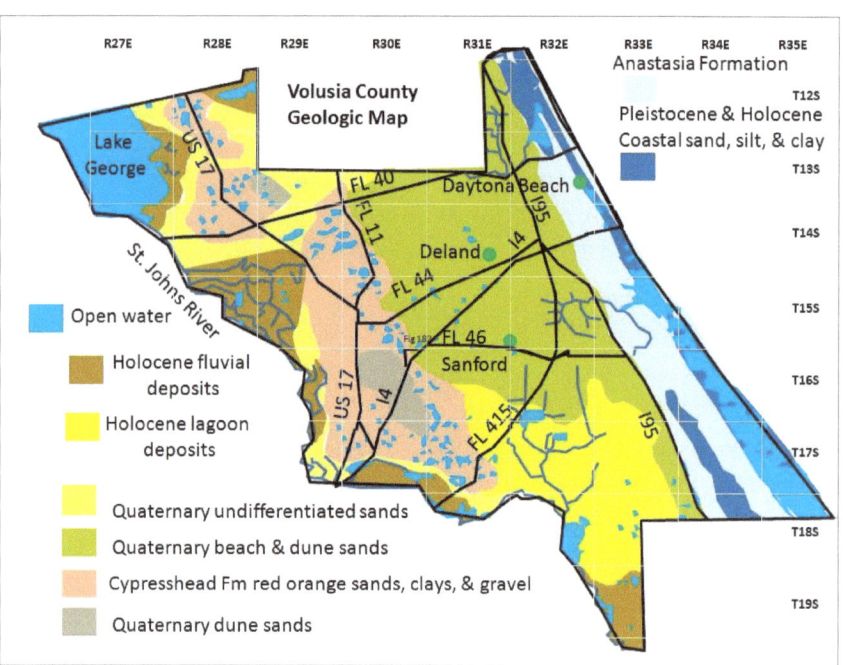

Figure 56. Volusia County geologic map. Revised from Florida Geological Survey Open File Map No. 8

Quaternary

Quaternary Holocene alluvial deposits (Qr)- Lowlands surrounding the St. Johns River and some of its tributaries in Volusia County occur as Holocene fluvial sediment alluvium composed of quartz sands, silt, clay, marl poorly consolidated, variably sandy, clayey, and shelly carbonates. Peat and other organic rich sediments are often present and is common below ten feet in elevation.

Quaternary Anastasia Formation (Qa)- occurs along the coast, on the barrier island, and the Atlantic Coastal Ridge on the mainland. This unit consists of variably lithified coquina and mixtures of sand and shell. It is well exposed along the Intracoastal Waterway near Ormond Beach and in pits along the Atlantic Coastal Ridge.

Quaternary Pleistocene/Holocene lagoons (Qph)- Along the present day coast, the associated lagoons and coastal rivers and streams are sediments referred to as undifferentiated Pleistocene and Holocene coastal deposits. The sediments are composed of sands, silts, and clays that sometimes contain varying percentages of organic matter. The sands may contain heavy minerals and are poorly to well sorted depending on the depositional environment and may contain varying amounts of fossils. The depositional environments include beach, marsh, and lagoonal sediments which rest unconformably on older, undifferentiated Quaternary sediments or Nashua Formation. Some occurrences of the Anastasia Formation may be included in the unit.

Quaternary beach & dune deposits (Qbd)- Beach ridges and dunes were deposited over much of the eastern half of the county. Quartz sands comprising the unit are generally fine to medium grained, moderately to well sorted and unfossiliferous. Sands may contain heavy minerals and deposits of organic material forming in swales between dunes and between beach ridges. The unit may rest unconformably on undifferentiated Quaternary sediments or Nashua Formation. The surface expression of this unit controls the drainage patterns and streams follow the swales, parallel to the coast until they encounter larger streams crossing perpendicular ridges.

Quaternary undifferentiated lagoonal sediments (Qul)- In southern central Volusia County these sediments are similar to Qu units and appear to have been deposited in lagoons landward of the coastal barrier islands.

Quaternary undifferentiated sediments (Qu)- This unit is present in western Volusia County composed of sands, clayey sands, and clays occasionally containing limited numbers of mollusk shells. Surficial conditions are often swampy and there are accumulations of organic matter. These sediments were deposited under marine conditions although some fluvial reworking occurred while possibly containing some Cypresshead Formation reworked sediments. This unit rests unconformably on the Nashua Formation or undifferentiated Pleistocene sediments. Beach and dune, and fluvial or lagoonal units may overlie this unit.

Tertiary

Tertiary Late Pliocene Cypresshead Formation (Tc)- consisting of quartz sands ranging from fine to very coarse, moderately to well sorted with common occurences of quartz gravel. Clay is commonly present in very minor amounts and is generally kaolinitic. Mica often occurs in minor percentages particularly in finer grained sediments. Colors range from reddish orange in exposed sections to olive gray in the subsurface. Faint, poorly preserved mollusk molds are present in exposures. The Cypresshead grades laterally down dip into the Nashua Formation, a variable shelly clayey sand. It unconformably overlies the Hawthorn Group and is overlain by sand of the Trail Ridge unconformably in western Clay County and thin surficial sands and soils elsewhere. The Cypresshead was deposited in a shallow nearshore marine setting.

South Peninsular Florida

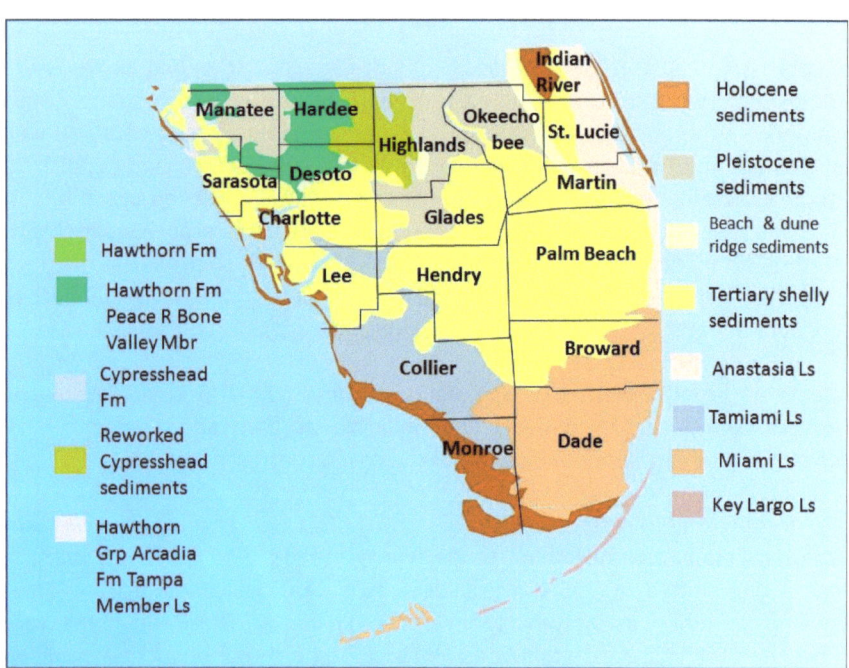

Index map of South Florida

Broward County

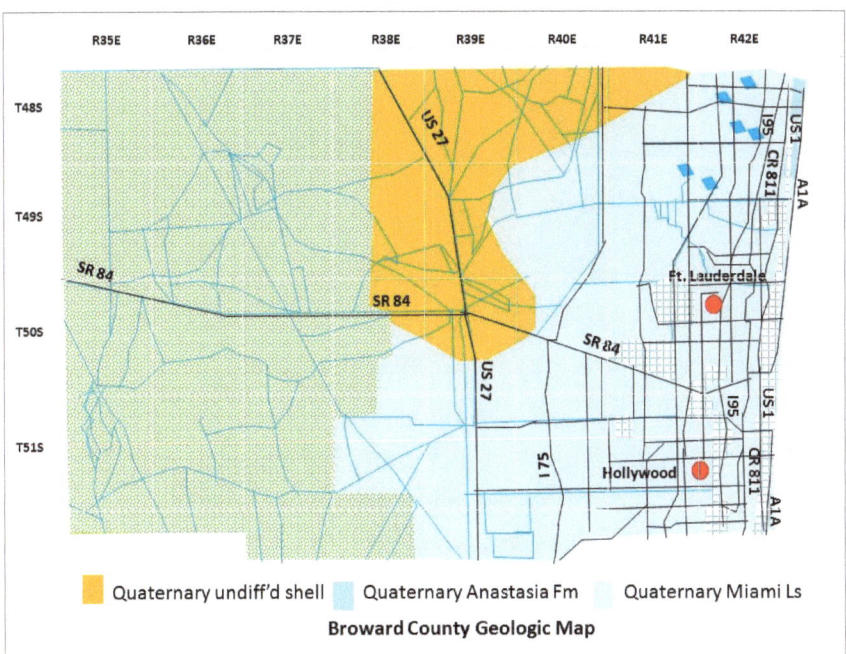

Figure 57. Broward County geologic map. Revised from Florida Geological Survey Open File Map No. 64.

Quaternary

Quaternary Anastasia Formation (Qa)- Variably lithified coquina of shells and sands and unlithified fossiliferous sand.

Quaternary Miami Limestone (Qm)- White to light gray limestone variably fossiliferous, oolitic, and pelletal. Variable percentages of quartz sand ranging to a sandy limestone to a calcareous sand.

Quaternary shell beds (Qsu)- Undifferentiated shell beds including sediments previously placed with the Caloosahatchie, Ft. Thompson, and Nashua Formations including the Pinecrest Beds.

Charlotte County

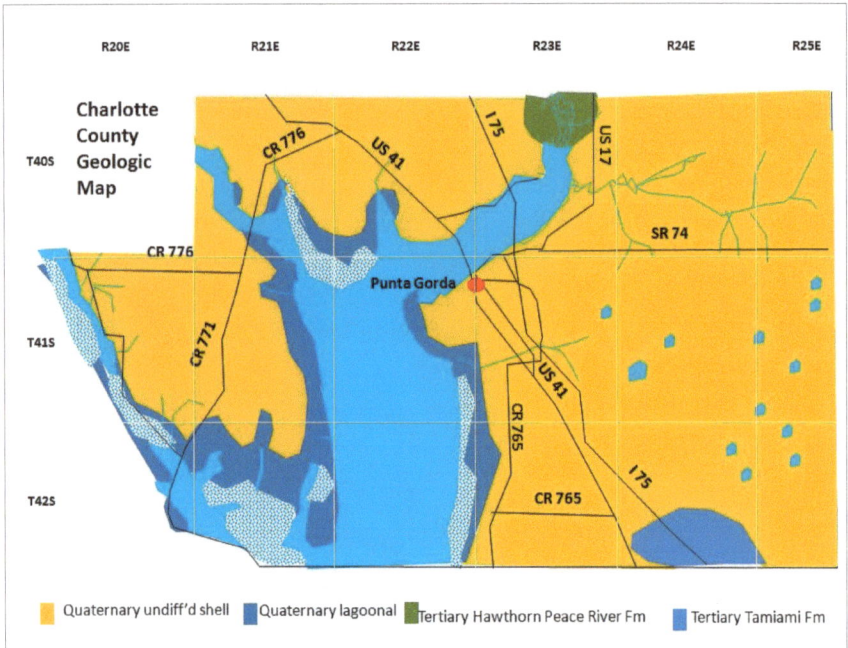

Figure 58. Charlotte County geologic map. Revised from Florida Geological Survey Open File Map No. 59.

Quaternary

Quaternary Holocene lagoonal (Qh)- Holocene sediments consisting of quartz sand with minor amounts of organic matter and clay associated with lagoonal deposits. Mostly beach & dune sands along present coastline with no formations recognized.

Quaternary shell beds (Qsu)- undifferentiated shell beds including sediments placed in the Caloosahatchie, Ft. Thompson, and Nashua Formation including the Pinecrest Beds.

Quaternary undifferentiated sands (Qu)- Undifferentiated surficial sands, clayey sands, clays, marls, and peat greater than 20 feet thick with no formations recognized.

Tertiary

Tertiary Hawthorn Group Peace River formation (Thpr)- consists of interbedded quartz sands, clays, and carbonates all of which are variably phosphatic.

Tertiary Tamiami Formation (Tt)- Sandy limestone, sands, clays, marls, with a variable phosphatic and fossil content. Buckingham marl member contains significant percentages of phosphate.

Collier County

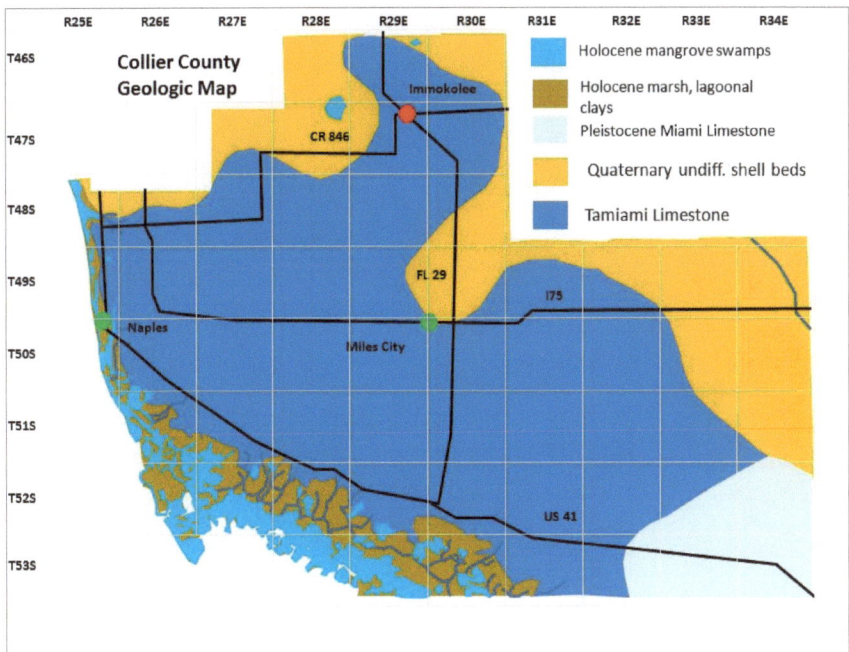

Figure 59. Collier County geologic map. Revised from Florida Geological Survey Open File Map No. 63.

Quaternary

Quaternary Holocene (Qh)- Holocene sediments consisting of quartz sand with minor amounts of organic matter and clay associated with lagoonal deposits. Mostly beach and dune sands along present coastline with no formations recognized.

Quaternary Miami Limestone (Qm)- White to light gray limestone variably fossiliferous, oolitic, and pelletal. Variable percentages of quartz sand some ranging from a sandy limestone to a calcareous quartz sand.

Quaternary shell beds (Qsu)- undifferentiated shell beds including sediments placed in the Caloosahatchie, Ft. Thompson, and Nashua Formation including the Pinecrest Beds.

Tertiary

Tertiary Tamiami Formation (Tt)- Sandy limestone, sands, clays, marls with a variable phosphate and fossil content. Often very fossiliferous. Buckingham marl member contains significant percentages of phosphate.

Dade County

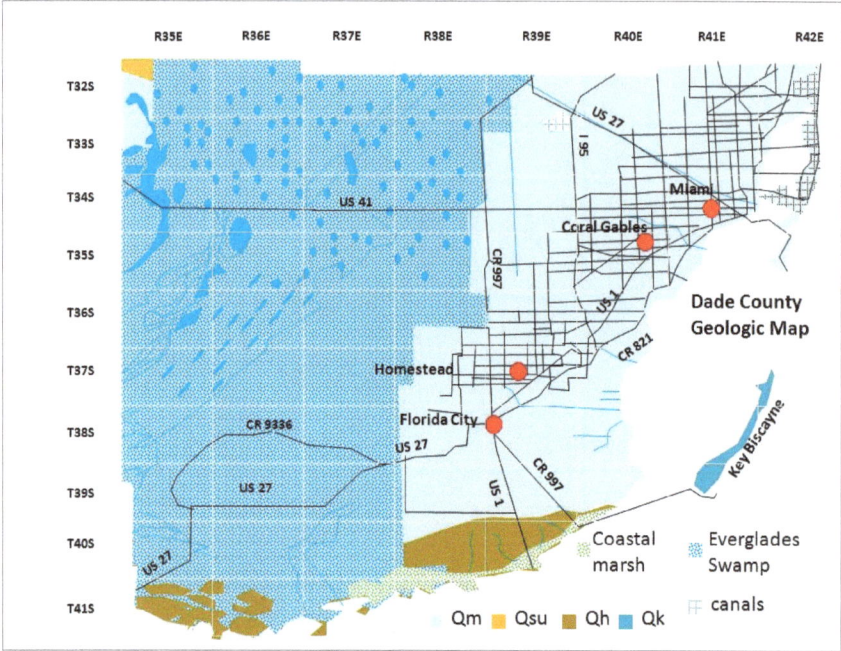

Figure 60. Dade County geologic map. Revised from Florida Geological Survey Open File Map No. 67.

Quaternary

Quaternary Holocene (Qh)- Holocene sediments consisting of quartz sand with minor amounts of organic matter and clay associated with lagoonal deposits. Mostly beach and dune sands along present coastline with no formations recognized.

Quaternary Key Largo Limestone (Qk)- White to gray highly fossiliferous coastline limestone with a fossil coral reef facies.

Quaternary Miami Limestone (Qm)- White to light gray limestone variably fossiliferous, oolitic, and pelletal. Variable percentages of quartz sand some ranging from a sandy limestone to a calcareous quartz sand.

Quaternary shell beds (Qsu)- undifferentiated shell beds including sediments placed in the Caloosahatchie, Ft. Thompson, and Nashua Formation including the Pinecrest Beds.

DeSoto County

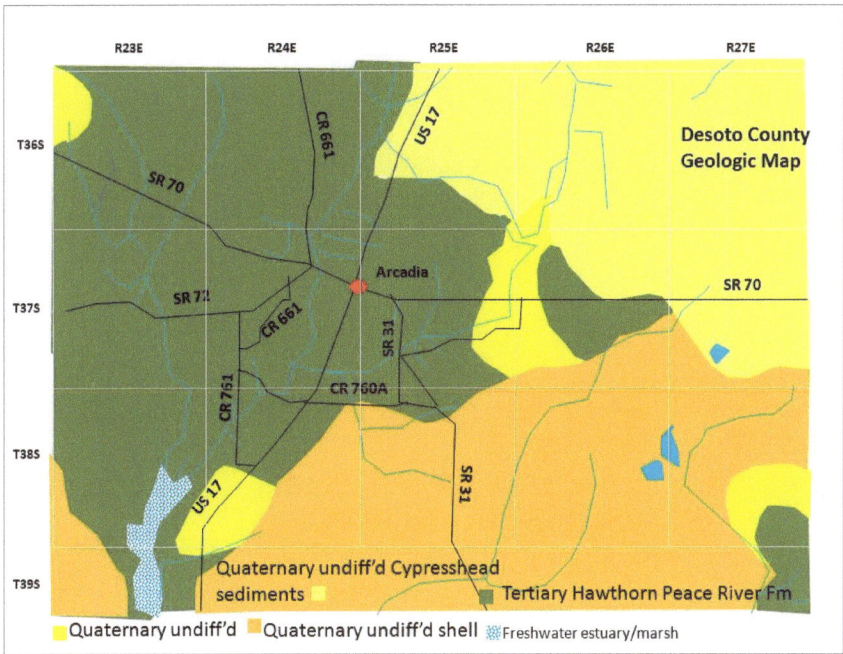

Figure 61. DeSoto County geologic map. Revised from Florida Geological Survey Open File Map No. 58.

Quaternary

Quaternary undifferentiated sands (Qu)- Undifferentiated surficial sands, clayey sands, clays, marls, and peat greater than 20 feet thick with no formations recognized.

Quaternary shell beds (Qsu)- undifferentiated shell beds including sediments placed in the Caloosahatchie, Ft. Thompson, and Nashua Formation including the Pinecrest Beds.

Quaternary undifferentiated Cypresshead (Quc)- unidfferentiated sands reworked from Cypresshead Formation with isolated occurences of Cypresshead Formation.

Tertiary

Tertiary Hawthorn Group Peace River formation (Thpr)- consists of interbedded quartz sands, clays, and carbonates all of which are variably phosphatic.

Glades County

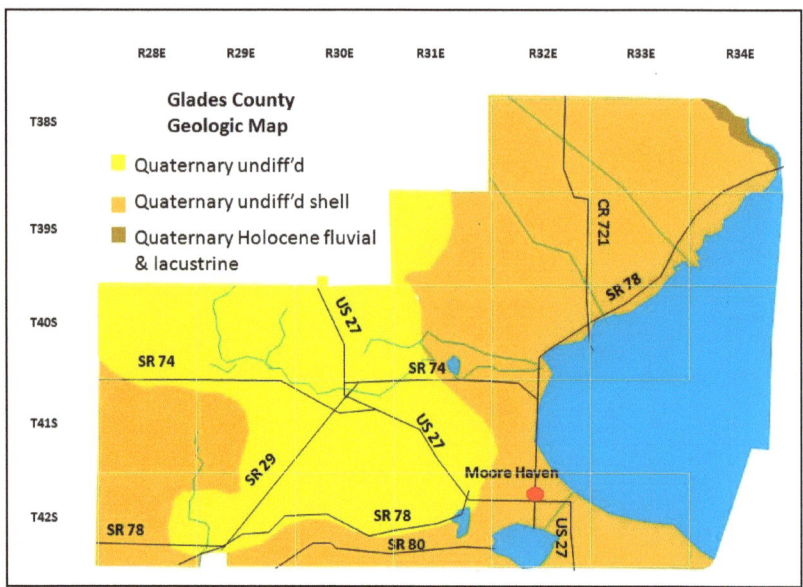

Figure 62. Glades County geologic map. Revised from Florida Geological Survey Open File Map No. 60.

Quaternary

Quaternary Holocene (Qr)- Holocene fluvial and lacustrine sands, clays, marls, and peats. No formation recognized. Use is restricted to peninsular areas only.

Quaternary shell beds (Qsu)- undifferentiated shell beds including sediments placed in the Caloosahatchie, Ft. Thompson, and Nashua Formation including the Pinecrest Beds.

Quaternary undifferentiated sands (Qu)- Undifferentiated surficial sands, clayey sands, clays, marls, and peat greater than 20 feet thick with no formations recognized.

Hardee County

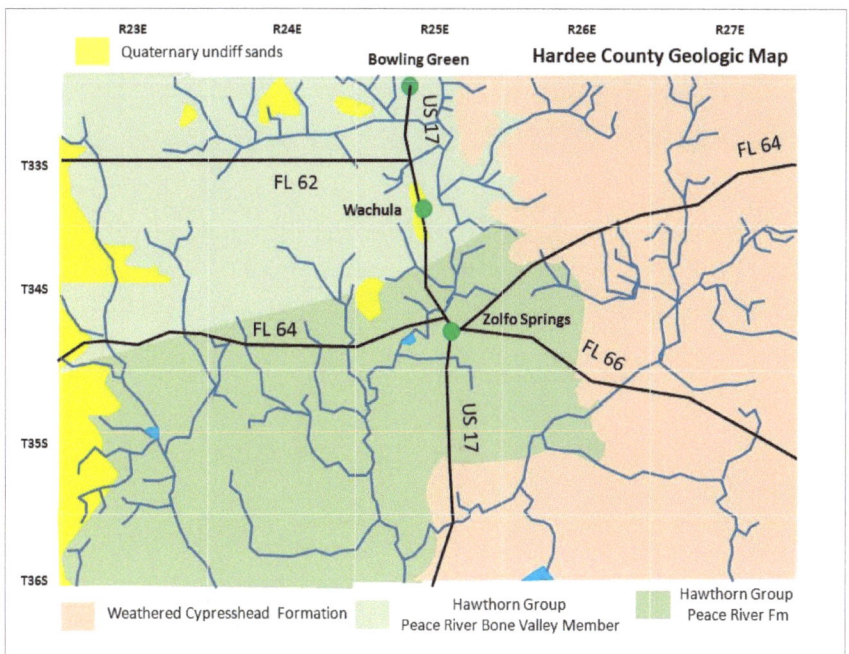

Figure 63. Hardee County geologic map. Revised from Florida Geological Survey Open File Map No. 51.

Quaternary

Quaternary undifferentiated sands (Qu)- Undifferentiated surficial sands, clayey sands, clays, marls, and peat greater than 20 feet thick with no formations recognized.

Quaternary undifferentiated Cypresshead (Quc)- unidfferentiated sands reworked from Cypresshead Formation with isolated occurences of Cypresshead Formation.

Tertiary

Tertiary Hawthorn Group Peace River Formation Bone Valley Member (Thpb)- Pebble or gravel sized phosphate fragments and sand sized phosphate grains in a matrix of quartz sand and clay. Percentages of the varous components are variable.

Tertiary Hawthorn Group Peace River formation (Thpr)- consists of interbedded quartz sands, clays, and carbonates all of which are variably phosphatic.

Hendry County

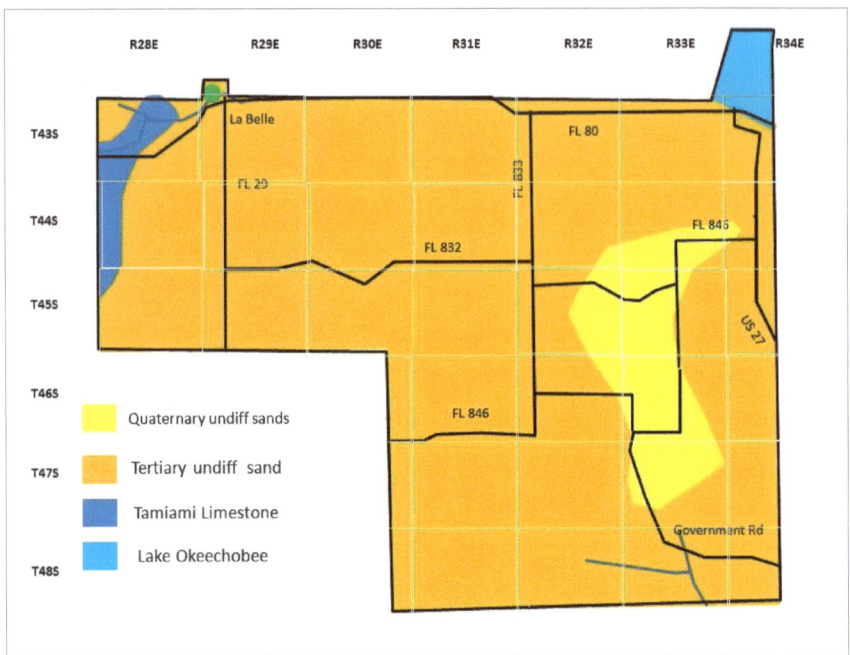

Figure 64. Hendry County geologic map. Revised from Florida Geological Survey Open File Map No. 62.

Quaternary

Quaternary shell beds (Qsu)- undifferentiated shell beds including sediments placed in the Caloosahatchie, Ft. Thompson, and Nashua Formation including the Pinecrest Beds.

Quaternary undifferentiated sands (Qu)- Undifferentiated surficial sands, clayey sands, clays, marls, and peat greater than 20 feet thick with no formations recognized.

Tertiary

Tertiary Tamiami Formation (Tt)- Sandy limestone, sands, clays, marls with a variable phosphate and fossil content. Often very fossiliferous. Buckingham marl member contains significant percentages of phosphate.

Highlands County

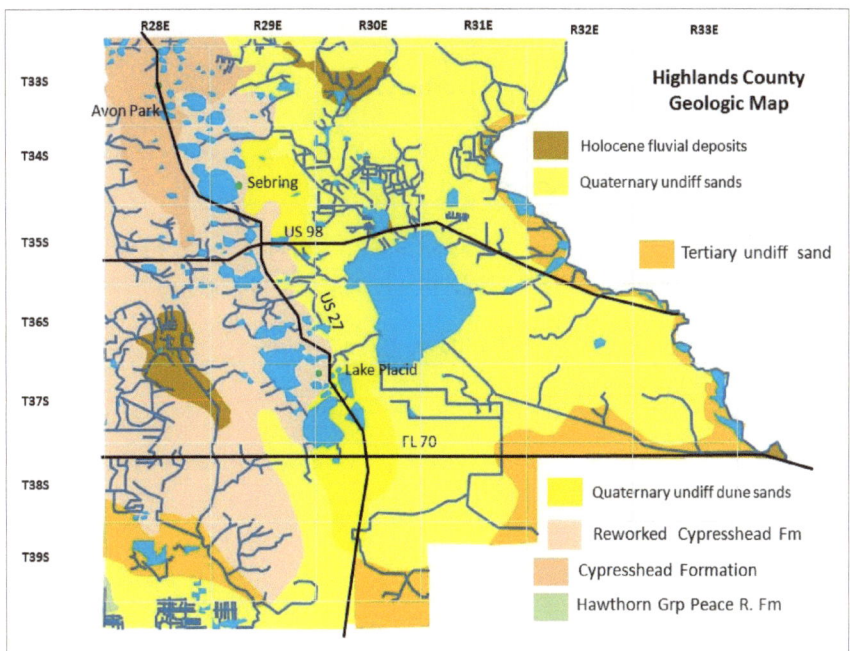

Figure 65. Highlands County geologic map. Revised from Florida Geological Survey Open File Map No. 52.

Quaternary

Quaternary Holocene (Qr)- Holocene fluvial and lacustrine sands, clays, marls, and peats. No formation recognized. Use is restricted to peninsular areas only.

Quaternary shell beds (Qsu)- undifferentiated shell beds including sediments placed in the Caloosahatchie, Ft. Thompson, and Nashua Formation including the Pinecrest Beds.

Quaternary undifferentiated sands (Qu)- Undifferentiated surficial sands, clayey sands, clays, marls, and peat greater than 20 feet thick with no formations recognized.

Quaternary undifferentiated Cypresshead (Quc)- undfferentiated sands reworked from Cypresshead Formation with isolated occurences of Cypresshead Formation.

Quaternary/Tertiary

Quaternary/Tertiary dunes (QTd)- Quartz sands with surface expression of dunes. A geomorphic unit on undifferentiated often quartz sands generally fine to medium grained.

Tertiary

Tertiary Cypresshead Formation (Tc)- Quartz sands with minor clay resting on top of Hawthorn Group sediments.

Tertiary Hawthorn Group Peace River formation (Thpr)- consists of interbedded quartz sands, clays, and carbonates all of which are variably phosphatic.

Indian River County

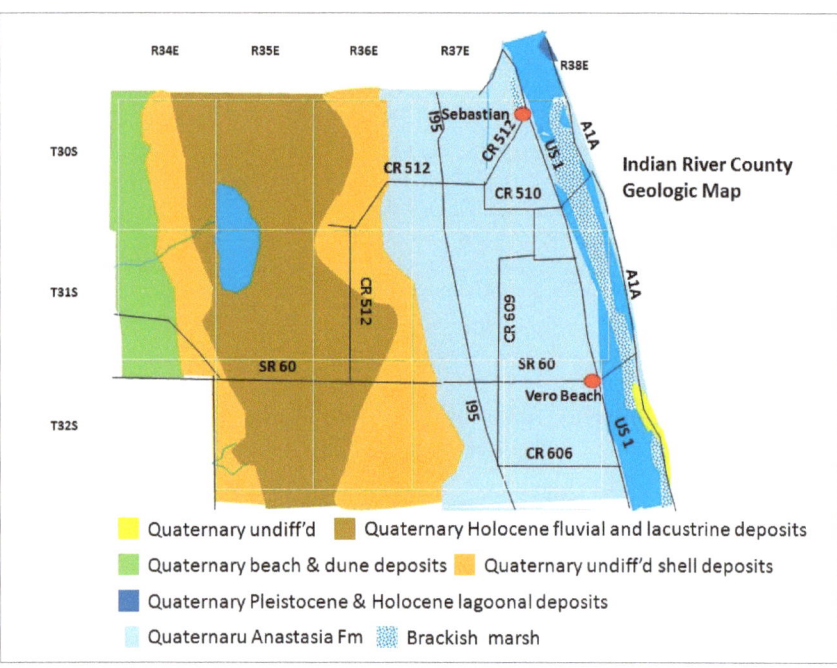

Figure 66. Indian River County geologic map. Revised from Florida Geological Survey Open File Map No. 55.

Quaternary

Quaternary Anastasia Formation (Qa)- Variably lithified coquina of shells and sands and unlithified fossiliferous sand.

Quaternary beach & dune deposits (Qbd)- Quartz sands with surface expressions of beach ridges and dunes. A geomorphic unit on undfferentiated clean quartz sands. Generally fine to medium grained with no formations recognized. May contain shell.

Quaternary Holocene (Qh)- Holocene sediments consisting of quartz sand with minor amounts of organic matter and clay associated with lagoonal deposits. Mostly beach and dune sands along present coastline with no formations recognized.

Quaternary Holocene (Qr)- Holocene fluvial and lacustrine sands, clays, marls, and peats. No formation recognized. Use is restricted to peninsular areas only.

Quaternary shell beds (Qsu)- undifferentiated shell beds including sediments placed in the Caloosahatchie, Ft. Thompson, and Nashua Formation including the Pinecrest Beds.

Quaternary undifferentiated sands (Qu)- Undifferentiated surficial sands, clayey sands, clays, marls, and peat greater than 20 feet thick with no formations recognized.

Lee County

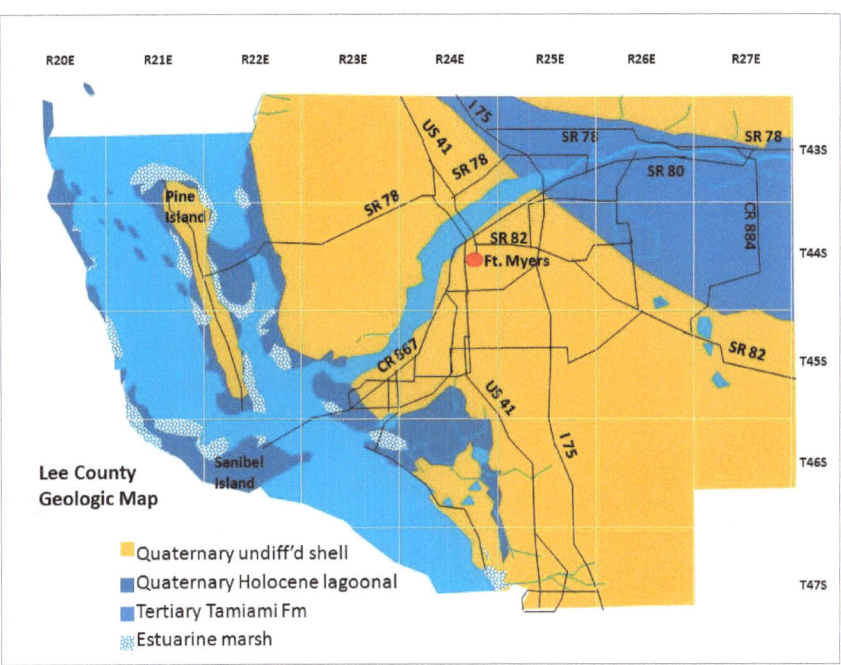

Figure 67. Lee County geologic map. Revised from Florida Geological Survey Open File Map No. 61.

Quaternary

Quaternary Holocene (Qh)- Holocene sediments consisting of quartz sand with minor amounts of organic matter and clay associated with lagoonal deposits. Mostly beach and dune sands along present coastline with no formations recognized.

Quaternary shell beds (Qsu)- undifferentiated shell beds including sediments placed in the Caloosahatchie, Ft. Thompson, and Nashua Formation including the Pinecrest Beds.

Tertiary

Tertiary Tamiami Formation (Tt)- Sandy limestone, sands, clays, marls with a variable phosphate and fossil content. Often very fossiliferous. Buckingham marl member contains significant percentages of phosphate.

Manatee County

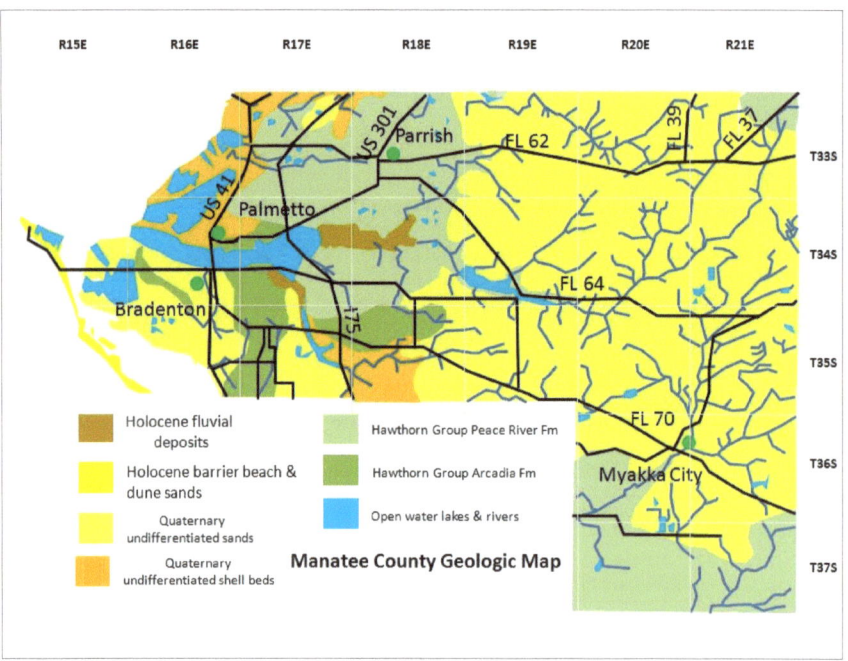

Figure 68. Manatee County geologic map. Revised from Florida Geological Survey Open File Map No. 50.

Quaternary

Quaternary Holocene (Qh)- Holocene sediments consisting of quartz sand with minor amounts of organic matter and clay associated with lagoonal deposits. Mostly beach and dune sands along present coastline with no formations recognized.

Quaternary Holocene (Qr)- Holocene fluvial and lacustrine sands, clays, marls, and peats. No formation recognized. Use is restricted to peninsular areas only.

Quaternary shell beds (Qsu)- undifferentiated shell beds including sediments placed in the Caloosahatchie, Ft. Thompson, and Nashua Formation including the Pinecrest Beds.

Quaternary undifferentiated sands (Qu)- Undifferentiated surficial sands, clayey sands, clays, marls, and peat greater than 20 feet thick with no formations recognized.

Tertiary

Tertiary Hawthorn Group Arcadia Formation Tampa Member (That)- Consists predominantly of limestone with subordinate dolostone, quartz sands, and clays. Limestone contains various quartz sand and clay but contains little or no phosphate.

Tertiary Hawthorn Group Peace River Formation Bone Valley Member (Thpb)- Pebble or gravel sized phosphate fragments and sand sized phosphate grains in a matrix of quartz sand and clay. Percentages of the varous components are variable.

Tertiary Hawthorn Group Peace River formation (Thpr)- consists of interbedded quartz sands, clays, and carbonates all of which are variably phosphatic.

Martin County

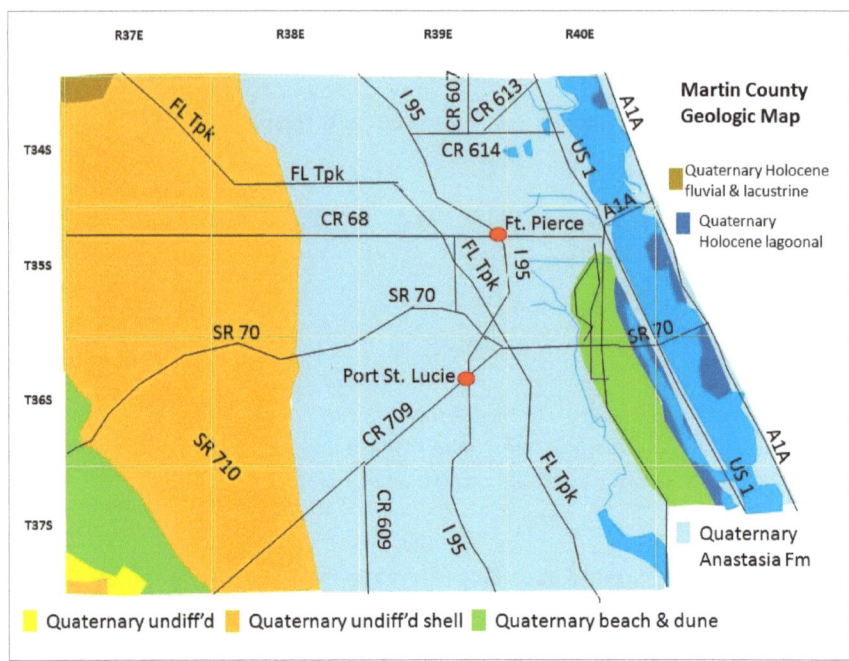

Figure 69. Martin County geologic map. Revised from Florida Geological Survey Open File Map No. 56.

Quaternary

Quaternary Anastasia Formation (Qa)- Variably lithified coquina of shells and sands and unlithified fossiliferous sand.

Quaternary beach & dune deposits (Qbd)- Quartz sands with surface expressions of beach ridges and dunes. A geomorphic unit on undfferentiated clean quartz sands. Generally fine to medium grained with no formations recognized. May contain shell.

Quaternary dunes (Qd)- Quartz sands with surface expressions of dunes. A geomorphic unit on undifferentiated often clean quartz sands generally fine to medium grained.

Quaternary Holocene (Qh)- Holocene sediments consisting of quartz sand with minor amounts of organic matter and clay associated with lagoonal deposits. Mostly beach and dune sands along present coastline with no formations recognized.

Quaternary shell beds (Qsu)- undifferentiated shell beds including sediments placed in the Caloosahatchie, Ft. Thompson, and Nashua Formation including the Pinecrest Beds.

Quaternary undifferentiated sands (Qu)- Undifferentiated surficial sands, clayey sands, clays, marls, and peat greater than 20 feet thick with no formations recognized.

Monroe County

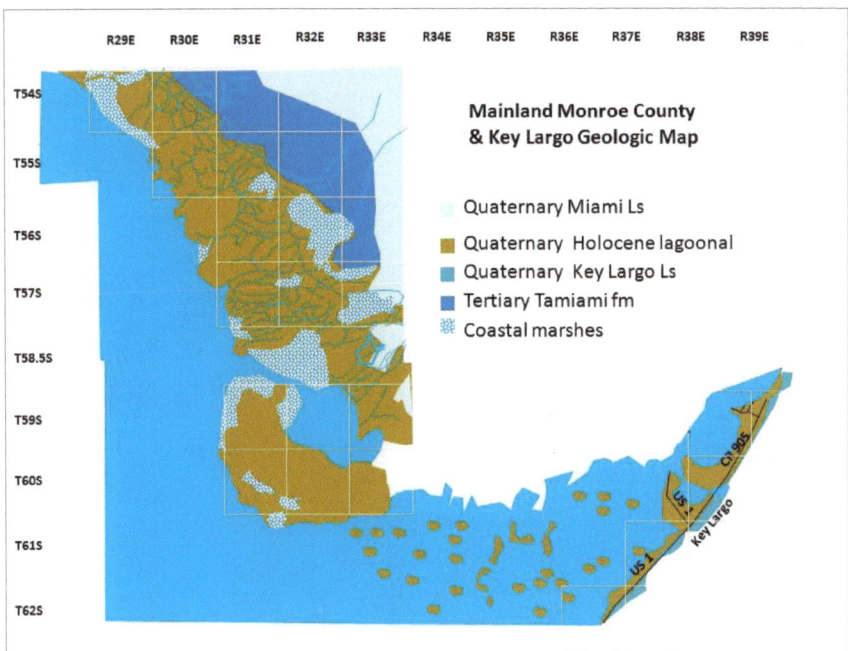

Figure 70. Monroe County Mainland geologic map. Revised from Florida Geological Survey Open File Map No. 66.

Quaternary

Quaternary Holocene (Qh)- Holocene sediments consisting of quartz sand with minor amounts of organic matter and clay associated with lagoonal deposits. Mostly beach and dune sands along present coastline with no formations recognized.

Quaternary Key Largo Limestone (Qk)- White to gray highly fossiliferous coastline limestone with a fossil coral reef facies.

Quaternary Miami Limestone (Qm)- White to light gray limestone variably fossiliferous, oolitic, and pelletal. Variable percentages of quartz sand some ranging from a sandy limestone to a calcareous quartz sand.

Tertiary

Tertiary Tamiami Formation (Tt)- Sandy limestone, sands, clays, marls with a variable phosphate and fossil content. Often very fossiliferous. Buckingham marl member contains significant percentages of phosphate.

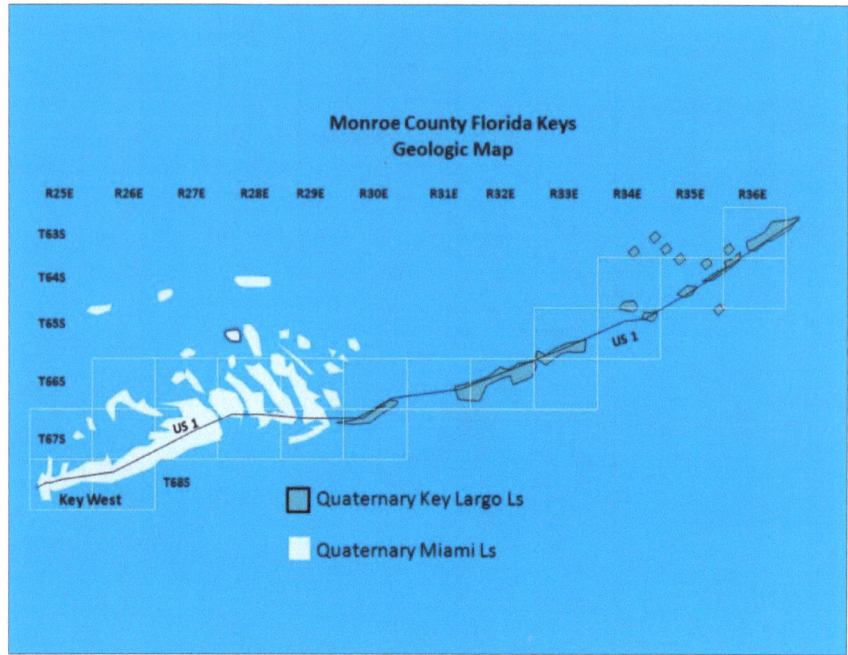

Figure 71. Monroe County Florida Keys geologic map. Revised from Florida Geological Survey Open File Map No. 66.

Quaternary

Quaternary Key Largo Limestone (Qk)- White to gray highly fossiliferous coastline limestone with a fossil coral reef facies.

Quaternary Miami Limestone (Qm)- White to light gray limestone variably fossiliferous, oolitic, and pelletal. Variable percentages of quartz sand some ranging from a sandy limestone to a calcareous quartz sand.

Okeechobee County

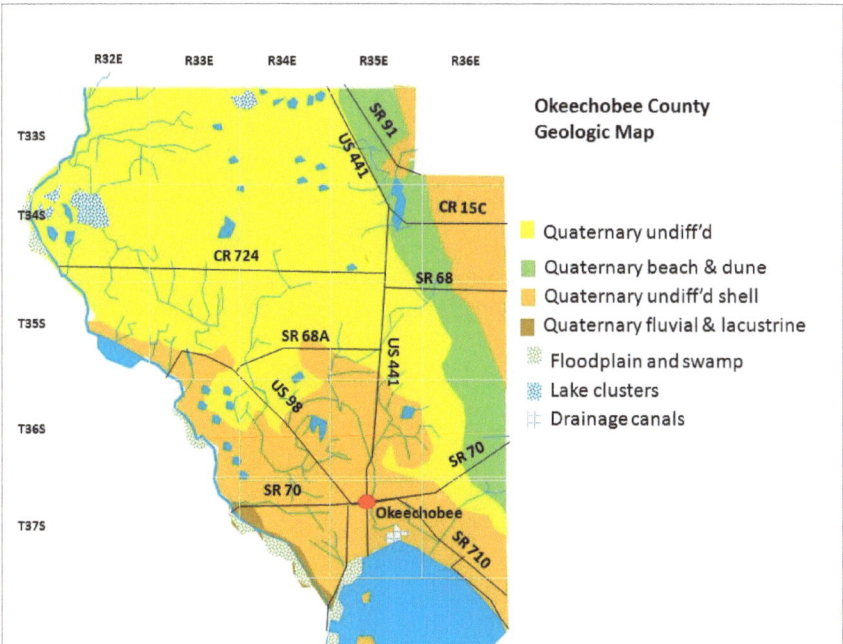

Figure 72. Okeechobee County geologic map. Revised from Florida Geological Survey Open File Map No. 54.

Quaternary

Quaternary beach & dune deposits (Qbd)- Quartz sands with surface expressions of beach ridges and dunes. A geomorphic unit on undfferentiated clean quartz sands. Generally fine to medium grained with no formations recognized. May contain shell.

Quaternary Holocene (Qr)- Holocene fluvial and lacustrine sands, clays, marls, and peats. No formation recognized. Use is restricted to peninsular areas only.

Quaternary shell beds (Qsu)- undifferentiated shell beds including sediments placed in the Caloosahatchie, Ft. Thompson, and Nashua Formation including the Pinecrest Beds.

Quaternary undifferentiated sands (Qu)- Undifferentiated surficial sands, clayey sands, clays, marls, and peat greater than 20 feet thick with no formations recognized.

Flood plain and swamp deposits – occurring along the Kissimmee River to the west along the floodplain and along the shorelines of Lake Okeechobee.

Lake clusters- appear in the northern part of the county and in the central western part of the county north of Lake Okeechobee.

Drainage canals occur along the central north, along the Kissimmee River, and around the edges of Lake Okeechobee.

Palm Beach County

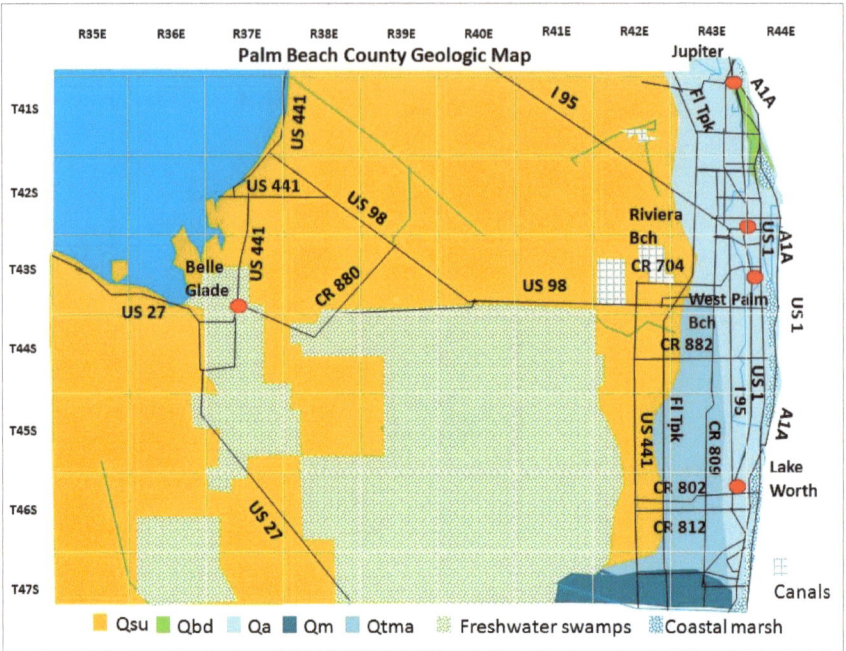

Figure 73. Palm Beach County geologic map. Revised from Florida Geological Survey Open File Map No. 65.

Quaternary

Quaternary Anastasia Formation (Qa)- Variably lithified coquina of shells and sands and unlithified fossiliferous sand.

Quaternary dunes (Qd)- Quartz sands with surface expressions of dunes. A geomorphic unit on undifferentiated often clean quartz sands generally fine to medium grained.

Quaternary Miami Limestone (Qm)- White to light gray limestone variably fossiliferous, oolitic, and pelletal. Variable percentages of quartz sand some ranging from a sandy limestone to a calcareous quartz sand.

Quaternary shell beds (Qsu)- undifferentiated shell beds including sediments placed in the Caloosahatchie, Ft. Thompson, and Nashua Formation including the Pinecrest Beds.

Quaternary transitional (Qtma)- Transitional between Miami Limestone and Anastasia Formation with some characteristics of each.

St. Lucie County

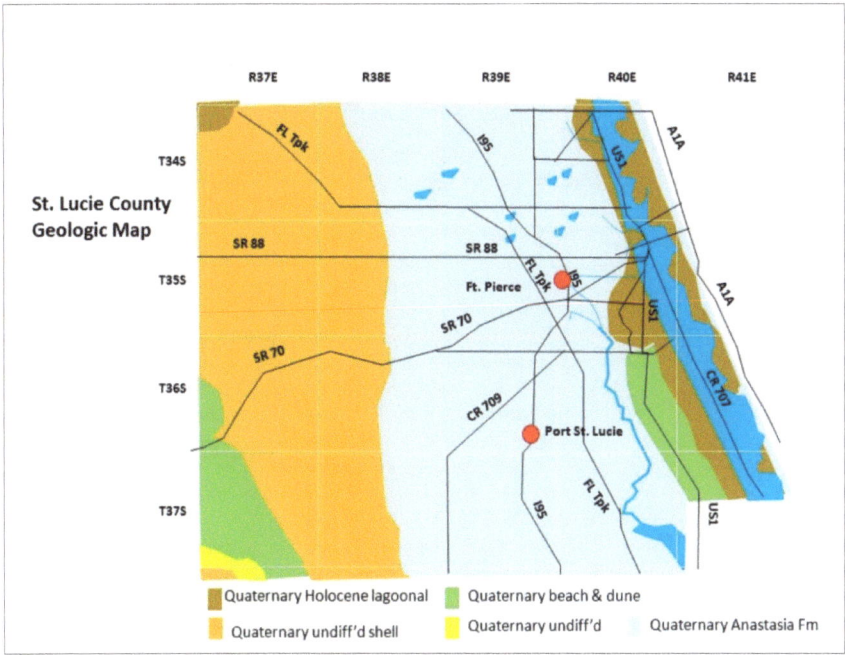

Figure 74. St. Lucie County geologic map. Revised from Florida Geological Survey Open File Map No. 53.

Quaternary

Quaternary Anastasia Formation (Qa)- Variably lithified coquina of shells and sands and unlithified fossiliferous sand.

Quaternary beach & dune deposits (Qbd)- Quartz sands with surface expressions of beach ridges and dunes. Generally fine to medium grained with no formations recognized. May contain shell.

Quaternary Holocene (Qh)- Holocene sediments consisting of quartz sand with minor amounts of organic matter and clay associated with lagoonal deposits. Mostly beach and dune sands along present coastline with no formations recognized.

Quaternary Holocene (Qr)- Holocene fluvial and lacustrine sands, clays, marls, and peats. No formation recognized. Use is restricted to peninsular areas only.

Quaternary shell beds (Qsu)- undifferentiated shell beds including sediments placed in the Caloosahatchie, Ft. Thompson, and Nashua Formation including the Pinecrest Beds.

Quaternary undifferentiated sands (Qu)- Undifferentiated surficial sands, clayey sands, clays, marls, and peat greater than 20 feet thick with no formations recognized.

Sarasota County

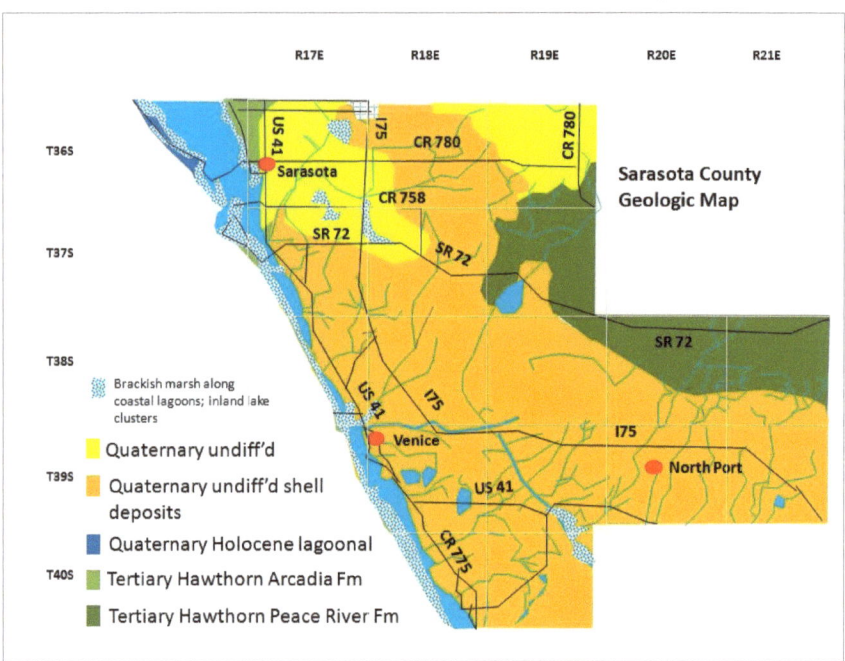

Figure 74. Sarasota County geologic map. Revised from Florida Geological Survey Open File Map No. 57.

Quaternary

Quaternary Holocene (Qh)- Holocene sediments consisting of quartz sand with minor amounts of organic matter and clay associated with lagoonal deposits. Mostly beach and dune sands along present coastline with no formations recognized.

Quaternary shell beds (Qsu)- undifferentiated shell beds including sediments placed in the Caloosahatchie, Ft. Thompson, and Nashua Formation including the Pinecrest Beds.

Quaternary undifferentiated sands (Qu)- Undifferentiated surficial sands, clayey sands, clays, marls, and peat greater than 20 feet thick with no formations recognized.

Tertiary

Tertiary Hawthorn Group Arcadia Formation (Tha)- undfferentiated Arcadia Formation consists of predominantly carbonates which are variably quartz and phosphatic sandy and clayey sediments. Sand and clay beds are often present and dolostone is generally the dominant carbonate component except in the Tampa Member.

Tertiary Hawthorn Group Peace River formation (Thpr)- consists of interbedded quartz sands, clays, and carbonates all of which are variably phosphatic.

References

Florida Geological Survey. (multiple dates). Open file Map Series 3 through 68 by various authors cited with each figure presented in this book.

Healy, H.G. 1979. Terraces and shorelines of Florida. US Geological Survey. Published by the Florida Department of Natural Resources, Bureau of Geology.

Randazzo, A.F., Jones, D.S., 1997. The geology of Florida. University Press of Florida, Gainesville, Florida.

White, W.A. 1970. The geomorphology of the Florida peninsula. Florida Geological Survey Bulletin No. 51. 164p.

www.ingramcontent.com/pod-product-compliance
Lightning Source LLC
Chambersburg PA
CBHW040219220526
45473CB00001B/53